Optimistic Storm:

Part 2: A Future Powered by Artificial Intelligence

Food to Feed All, Health for All, End Drudgery

and Restore our Environment

> *"Centuries pass when nothing happens, and then in a few short years events of such importance take place that the world is never the same again."*
>
> 1982 David Eddings.

Hugh Butler, PhD

ISBN: 978-0-9941733-4-8 Paperback
ISBN: 978-0-9941733-5-5 E-book

Edition 2 June 2026 -Updated

Structure of Book

These books and chapters are interconnected and we cannot talk about one without referencing multiple other subjects. Electricity is foundational. For the first time since humans walked on the earth, we no longer need to burn molecules for energy.

- Part 1 explores electricity disruption and the end of the fossil fuel era. Electrifying all energy cascades into transportation and stops the emissions that cause climate change. The electrification of energy changes our society.
- Part 2 explains how AI is about to transform food, health, agriculture, transport and labour.
- Part 3 concludes nuclear, hydrogen, carbon capture and storage and direct air capture distract us from our energy and AI-enabled future.

In Part 2 we focus on AI and the remarkable changes in abundance it will bring to knowledge work, physical labour, health, food, transport, and agriculture.

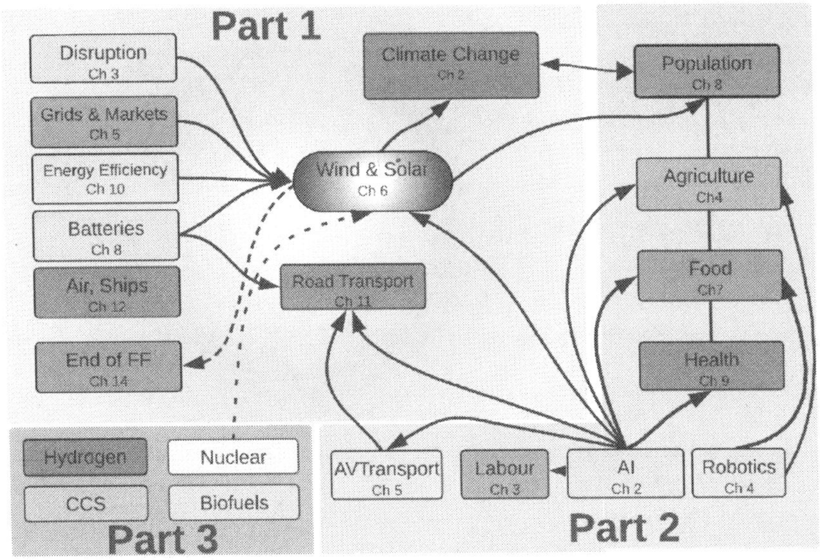

This 2nd edition includes feedback and comments for clarity

Table of Contents

1 Introduction

To be covered in the Introduction:

- The climate crisis demands we reduce emissions. (Part 1)
- The "carrying capacity" of the world (later chapter in Part 2) drives urgency to expand clean electricity and reduce the destruction of our environment.
- Review the 5 major disruptions in energy, labour, food, health, and transport to understand the future.
- Explore the policy failures and political murkiness behind these technologies and changes.
- Meet Ada and Tracey – two lives on opposite sides of the energy and wealth divide, who guide us through the book.
- Why people change: they make money, they lose money, or the law forces them to – the three drivers behind every disruption.

In Part 1 of Optimistic Storm, we looked at the disruption of energy through the eyes of Ada, a young sub-Saharan woman. She has a child slung on her back and two others at her side, and dreads going out into the blazing sun to walk five km to gather some wood to cook for her family. In contrast, Tracey, a middle-class person from a Western society, lives with modern amenities. Disruption and change are happening to both, but in quite different ways.

1.1 Disruptions Change Lives

In Part 2, we will continue to explore through the eyes of Ada and Tracey how they will be affected by the upcoming changes. These changes are possible by leveraging advances in artificial intelligence. Those advances in AI are cascading into multiple different areas of our lives, and because the AI is at such an early stage, it is extremely difficult to predict exactly where AI will take us in these diverse areas of energy, health, food, labour, agriculture and transport. Some predict a super abundance in those areas as energy and labour, the two pillars of our society, trend towards zero cost.

1. The end of labour and toil as we currently know it, along with a massive influx of humanoid robots.
2. Advances in Health, particularly with the seven big "omics of the genome, exome, epigenome, transcriptome, proteome, microbiome and metabiome.
3. Disruption of food and agriculture using precision fermentation and robotics.
4. The regreening of the planet as a consequence of changing farming practices and land use globally.
5. When travel becomes effortless.

We will explore these in separate chapters as the underlying technology is applicable across various uses. Moreover, we must understand some fundamental aspects of the human population to comprehend how these disruptions could unfold in our society. Similar to the energy disruption, advances in health, food and labour will change both the advantaged and the disadvantaged. Social upheaval is inevitable, but being prepared could enable a smoother transition if politicians and social belief systems support the changes. Some of you who read these changes outlined in this book will think this is science fiction. Some concepts are challenging.

One takeaway is that these changes are not decades away but are in commercial use today. All of them. No new technology needs to be invented for this to happen. We are simply expanding the use. Changes in the energy transformation are now visible to all. AI for drug development are already being used. AI for personal productivity are in use. What will it be like in 2030? Two decades by 2045? Will there be any job that a machine cannot do? How fast will the change be in these different areas? They are real, progressing at breakneck speed and will challenge us, our children and grandchildren and our society.

1.2 Energy and Energy Inequality

"Ada", a Nigerian name for first daughter, survives on just 1% of the world's resources. Another 800 million people like her are in energy poverty. Distressingly, many of those are subjected to violence, and for over 117 million, genocide, slavery and war. "Tracey," a middle-class, middle-aged woman who could be from any Western country, is wondering how she will pay her medical, energy, or insurance bill, which has just increased by 15% and has doubled in the past five years. Tracey is an example for those billion "wealthier" cohorts who own and use 71% of the world's resources.

We can be very optimistic about the end of the fossil fuel era. Electricity generated from renewable sources now surpasses electricity from coal. The costs of wholesale electricity in Australia are down 10% in 2026, back to levels before the price increases following the Ukraine war. It is no longer about whether we will have disruption, but about how quickly it will come. It is about 5% of the journey to change from polluting molecular energy to harvesting electricity. We're changing from consuming 200 years' worth of stored carbon to using just 1.5 hours of sunlight. Technology enables 66,000 TWh of electricity to displace 166,000 TWh of fossil fuel as we electrify everything - from heating, cooling, and transportation. Now, over 40% of electricity generated is clean, with 7% from solar and 8% from wind. Every 3 years, the amount of solar doubles. By 2027, 14% will be solar and by 2030, 30%. We can see how personal public transport using robotic electric vehicles will bring to an abrupt end the era of 1.4 billion internal combustion engine vehicles. Every AI-powered EV robotaxi will displace 10 cars. That will eat into the oil burnt.

Energy security through renewable energy will end fossil-fuel geopolitics, a plague on society for the past two hundred years. Just 24 countries lord it over 150 countries using their geopolitical advantage of fossil fuels. Wind and solar empower nations and restore their energy sources and sovereignty. No doubt there will be other geopolitical and religious reasons for conflict, but energy will no longer be the primary cause. Countries will find another reason for conflict.

Fossil fuels created energy poverty, and they cannot end it. Moreover, we can see a pathway to end energy poverty for Ada and that cohort of 800 million people. A Canadian uses 120 MWh per person of electricity a year; most Western societies use about 60 MWh; and Ada uses a fraction of that at just 120kWh. Can we dream of everyone having access to 200MWh, or of energy superabundance? Ada in the Global South and Tracey in the West will both reap the rewards. Why is this change not faster? As we see in the Population Chapter, some of the reasons behind this wealth inequality are that a billion people live on more than $100 per day, and a billion live on less than $1 per day.

1.3 Why People Change

Dr Adam Dorr of RethinkX, in his book Brighter: Optimism, Progress, and the Future of Environmentalism, [1] laid out the premise that prosperity will help drive change. A business colleague once told me people only change for one of three reasons, these being:

- They make money. i.e., There is a benefit, direct or indirect.
- They lose money. i.e., there is a cost, direct or indirect.
- It's the law. Politics intervenes and can accelerate or slow down change.

In each of these areas, we will discuss who makes or loses money and the implications for Ada and Tracey. For a fuller understanding of change and innovation, we deal with it in detail in Section 3 in Part 1.

1.4 Five Disruptions

We discussed in detail in Part 1, that a particular type of innovation creates what Tony Seba calls a "Phase Change Disruption". Phase-change disruptions create entirely new, much larger systems with new properties, new business models, and new metrics. It is not a one-to-one replacement. Phase change disruption is integral in:

- Energy (see section 4 in Part 1).
- Transportation (see section 7 in Part 1).
- Labour.
- Health.

- Food and Agriculture.

It is impossible to know in advance the full details of how systems will differ from those we have today. A key feature of this phase-change disruption is that the marginal cost of energy, transportation, labour, health, and food will rapidly approach zero. Ark Invest CEO Cathie Wood discusses 14 technology disruptions as core to the five societal disruptions. Tony Seba of RethinkX identifies eight technologies that drive three societal disruptions. Ark adds a number of financial disruptions, and RethinkX adds battery and wind.

1.5 Phase Change Disruption

Disruption is not a straight line. We don't always see a one-to-one input-to-output relationship in the world, as Prof Lee Fennell points out. A vending machine is very linear. You put in some money, and when you have put in enough, you get something out – a candy bar. [2] Some things work that way – predictably. Unless the machine jams!

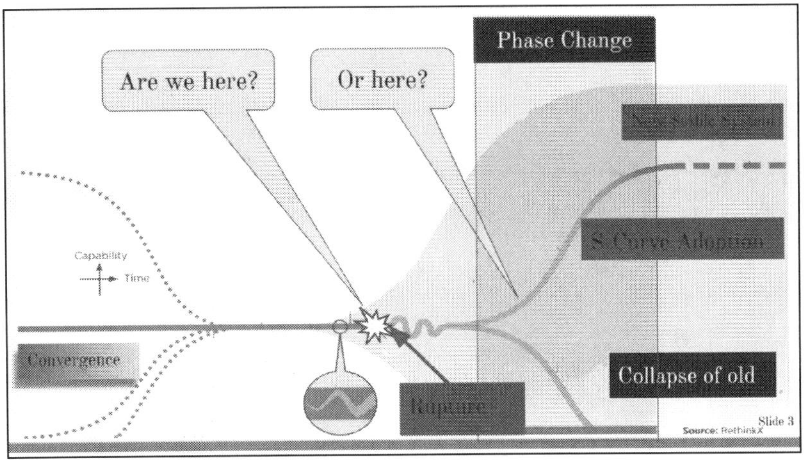

Figure 1 Phase Change Disruption and Rupture Points

Figure 1 shows that phase change disruption depends on multiple technologies coming together. If we consider an electric car – it needed good batteries and it was not the auto industry that developed good batteries, but music players and mobile phones. During phase-change disruptions, it can be difficult to determine when that rupture point occurs. It is also difficult to predict where on that path to a new stable system we are, or when the old system collapses.

Constructing a bridge is a non-linear project, but it is reasonably predictable. We receive almost zero return until all the inputs are in place and the bridge is fully constructed and open to traffic. A bridge is still relatively straightforward, whereas larger, more complex projects such as hydro and nuclear have their own set of cost and time blowouts, as Prof. Bent Flyvbjerg points out. [3]. AI, labour, food, and health are non-linear but also unpredictable. They may be more like an "S" curve. With certain technologies, we may experience very small or no returns initially but achieve huge returns later, or, conversely, have good returns early on, but the potential is never fully realised. The timing of when change occurs is the most important issue.

1.6 Change Can Be Fast

Change can be fast. When you hear commenters say the change from AI will be slow, refer them back to this image.

New York City, 1900 New York City, 1913

Figure 2 NY Easter Parade 1900-1913 One car then one horse

Figure 2 from Tony Seba [4] shows the Easter parade in New York – just 13 years apart. Not only did it move from spot the car in 1900 to spot the horse in 1913 but in that time a whole system including petrol refineries, distribution and logistics was built. When you have a compelling driver, change is fast. What became of the horse? The pet food industry was born, and our pets ate them.

In following chapters we will look at the impact of AI on various other disciplines. For example, with autonomous driving, it would be tempting to say that it will take another 2 decades for it to move from trial to full production, given that it has taken 2 decades to move from concept to 60 cities. However, understanding phase change disruption and rapid change, we can reasonably expect to see that technology will accelerate as AI improves.

1.7 AI for Labour for Services

Four out of five Australian workers are employed in services, or "white-collar" sector, reflecting the shift from manual and routine work to non-routine and so-called "cognitive" work. We sometimes call them knowledge workers. [5] AI will augment or displace jobs in this white-collar service industry. We will explore the social and economic ramifications of AI in labour, including the displacement of workers and the implications for GDP with an unlimited, low-cost workforce. Using AI could mean both white- and blue-collar jobs can be done onshore, rather than the past 50 years of sending service jobs and manufacturing offshore. We will explore how a country such as the Philippines, whose workforce in the offshoring of service jobs may need to adapt, and the profound societal changes that AI will bring. Ada will not have a call centre for employment in Sub-Saharan Africa when energy poverty is addressed. Will Tracey have her job as an administrator in a local business in 20 years?

1.8 AI for Robotics and Machines

AI and robotic systems free people from doing tasks that are dirty, dangerous or drudgery. In the jobs that do physical work, or the "blue-collar" sector, humanoid robotics will replace workers and, powered by AI, will replace physical jobs. RethinkX says humanoid robot disruption is here now. There is no stopping it. With over 50 companies already performing learning tasks in factories, robots will be doing real tasks in real factories by the end of 2025. Just as the car replaced the horse in 1905, humans are now the horses and will be replaced first in tasks and then in work.

My optimistic projection is that by 2030, we will have a few hundred million humanoid robots in businesses worldwide. Our upmarket friends and those with disabilities will have them in their households. Robots will replace cars as status symbols. By the mid-2030s, the number of robots could exceed a billion. Just ten years away! Expect more robots than humans within 20 years. Global geopolitical arrangements that promote the manufacture of goods in low-cost labour countries can be brought back to local manufacture. Countries relying on cheap labour for wealth are in for a major shock. Furthermore, robots will reduce the demand for the 60 million modern slaves as robots become cheaper labour than humans. Ada won't see it for some time, but it will come to her country because robots will be cheaper than local labour. As humanoid robots approach a cost of $0.50 per hour, establishing businesses in low-wage countries will be a thing of the past, not the future. The staff in Tracey's place of employment will be mostly laid off, with only a few managerial or specialist staff left.

Yet most governments fail to see this looming impact. Russia and a few other countries continue to encourage their populations to have more babies to increase the workforce. There won't be those jobs for them when they reach working age in 18 years. Tracey will see it first at the local hospital or supermarket. The implications for our societies are almost beyond our comprehension. Most critical will be to have a framework for how the generated wealth flows. Does it flow to the few existing billionaires, or to society in general? We need to consider these issues today, not when we are even more dominated by tech billionaires and their greed.

1.9 Health Care

Our definition of healthcare is more accurately described as "sick care". Doctors and hospitals are set up to treat sick people. Not to prevent people from getting sick. There are now more people dying from lifestyle diseases from obesity than from starvation. There are treatments for most bacterial and viral infections, yet people in countries with poor health care systems continue to die. Health spending accounts for 11% of global GDP or $9 trillion annually, so anything to make it more cost-effective or serve more people is welcome.

Advances in genomics and health care have been slow, but we are on the cusp of multi-omics moving from the lab to the wider population. Previously, the majority of the global increase in life expectancy was due to the control of infectious diseases using vaccines and antibiotics. Treatment for cancer or heart disease extends longevity by less than 10 years, although there are notable exceptions. What is the projection for the next decade? We can be very confident that advances in AI, robotics, and genomics will be very disruptive.

This is a challenge in an industry that is extremely conservative, controlled by a few very large companies, and whose medical profession is averse to any change. There are two areas where AI can make huge changes to improve health.

- Diagnosis. Recent studies of using AI and genomics show that we can detect and treat cancers and heart disease before any symptoms. We will see an increasing improvement of diagnosis, although we will see the regulatory barriers continue to hinder wider adoption. Could we see advances in diagnostics made available to the 6.5 billion people who lack access to quality medical diagnosis?
- Treatment. Some futurists are very bullish about the costs of treatment. We are already seeing that the development of drug therapies has reduced from $2.1 billion and 13 years for a new drug to the market, fall by 12 times to less than $700 million and 6 years. We have seen AI scour through all the open source papers and clinical trials and found existing, registered drugs that fit medical conditions that do not have existing products. The cost of drugs in the health system is less than 2% of the total health expenditure and providing treatments to previously untreatable diseases will improve hundreds of millions of lives.

Will the advances in health care be available to Ada as well as Tracey, or just the wealthiest cohorts? Ada already has a mobile phone to aid diagnosis, but what she does not have is access to any of the first-world drugs. And what of the "*alternative or complementary health care*" industry, a $180 billion market?

My optimistic projection is that by 2030, we will see the rise of more personalised, localised health systems. How will politics stifle adoption? We will explore this in greater detail in the Health Chapter.

1.10 Agriculture

Expansion of agriculture is the major cause of environmental destruction. Land clearing in Brazil and Australia, wholesale harvesting in the ocean, and the production of food and clothing have decimated habitats and species. Over 40% of food grown is used to feed animals, even as the number of starving people has dropped from 20% to less than 10% over the past three decades. In the UK and the USA, over 20% of animals killed for meat feed domestic pets. The next time Tracey feeds her furball cat that latest pet food, think of how many animals have been killed for that tasty morsel. Meanwhile, we know that protein shortages are a real problem for Ada and more than 3 billion people.

An even worse travesty is that 40% of the corn in the USA and Brazil, and palm oil in the tropics, is burnt in vehicles with E10 petrol or diesel. That burning of ethanol also increases CO_2 emissions. The various governments provide billions of subsidies to the various agricultural corporations and we look at this in more detail in Part 3 biofuel. An acre of land growing corn /ethanol would propel a car 15,000 miles. An acre of solar panels would power an EV over 900,000 miles and also enable crops to be grown underneath or alongside. In metric terms, a hectare of corn provides 60,000km for an internal combustion engine; a hectare of solar provides 3.6 million km for an EV. We have been encouraged to eat less meat and buy fewer clothes as an individual contribution to curbing agricultural expansion and rising emissions, but large-scale wind and solar will bring about change much faster, without any adverse impact on lifestyle.

1.11 Food Disruption

Many will be aware of the plant-based foods movement over the past decade, but that is not where future disruption will come from. The genetic precision fermentation revolution now underway is enabled by genomics and AI. Crops feed animals (dairy cows, beef cattle, chickens for eggs or meat and pigs). Animals are grown and then slaughtered for meat or, in some cases, products, including egg laying or fibre. Precision fermentation bypasses this crop-animal-slaughter process. The new process will be:

- Grow starch crops.
- Use genetically designed yeasts or bacteria in a container.
- The bugs convert starch to protein or fats.
- Harvest the product.
- It turns a 3% efficient process into a 95% efficient process.

Few recognise the advances and expansion of this technology, although companies globally are building out these "craft breweries" or precision fermentation facilities. The cost of protein will be 5 times cheaper than existing animal proteins by 2030, and 10 times cheaper than 2035. Eventually, proteins will be cheaper than sugar.

The changes in food production will lead to an outcome for all agriculture and food system. Changes driven by the electrification of transport will spell a huge reduction in biofuel industry and the growing of crops for energy. Precision fermentation will cause major changes to the dairy and livestock industry. Some of the probable outcomes will be:

- A land area the size of the USA, China and Australia will be retired from farming and can revert to forests and pastures.
- The increase in natural grasslands and forests could bring CO_2 levels back to longer-term trends, we get wilding of species, and the elimination of commercial fishing of oceans.
- Even without active reforestation, the passive reforestation of land through the process of natural recovery will capture and store carbon equivalent to up to 20% of today's global emissions.
- Food, locally made, removes sovereignty risk and the costs of transportation.

There is already a growing awareness of the impact on farming, and as the fossil fuel industry combats renewable energy, expect to see a similar strident opposition by the agricultural sector. A change is underway. This change will not affect Ada because for most of the Global South farms are small, less than 1ha. However this will have a massive impact on Tracey and her supply food as only dozen or so companies already provide 90% of Tracey's food. We will explore more in this Part 2.

1.12 Innovation & Global Wealth

Ark Invest has an interesting take on the impact of innovation on global wealth.[6] Prior to the advent and proliferation of writing, the global economy was anaemic, with growth rates below 0.04% over the 100,000 years leading up to the rise of Rome. In the first 1,000 years AD, annual growth accelerated nearly fourfold to 0.14%. Then, plough technology and crop rotation strategies led to a population boom, more than doubling annual growth to 0.3% through 1500.

The printing press and the steam engine bookended the first industrial revolution, doubling growth again to 0.6% per year through 1900. Thereafter, electrification, telephony, the internal combustion engine, and digital computation and connectivity quintupled real economic growth to 3% at a compound annual rate, pushing global gross domestic product (GDP) to $107 trillion. As a result, global real per capita GDP has increased nearly 7-fold since 1900 from less than $2,000 to more than $13,400 in 2023.

We see the flow-on effects in population growth, in less infant mortality, in the development of education and health in most countries. While Tracey is complaining about the cost of living and immigration and how things are no longer the way she remembers them, Ada will be grateful for the children's vaccine and mobile phone that tells her of a better life, sometime.

1.13 Conclusion of Introduction

This journey is not without its challenges. There will be obstacles to overcome, and the path forward will require collective action and sustained commitment from individuals, businesses, and policymakers alike. But within these pages, you'll find real-world examples, inspiring stories, and a deeper understanding of the extraordinary advancements already happening.

As we discovered in Part 1, disruptions in renewable energy and batteries are the beginning of the end of the fossil fuel era. By when? Here are just some of the characteristics of that disruption.

- There are limits to the rate of change, simply because it takes time for people to assemble solar panels and batteries in the real world.
- Pakistan showed how it could be done. It doubled its electricity supply in just 1 year because individuals installed their own solar. Governments have to get out the way, and speed up installation, not slow it down.
- Australia installed 400,000 batteries with a capacity of over 10GWh in just 9 months. That is about 10% of the battery capacity needed for Australia to be 95% renewable electricity. By 2030 when the program stops, it is expected to be about 50%.

- EVs are the nail in the coffin of oil. EV sales are 1 in 4 new cars globally and in some countries 100%. The model is there. Will other countries adopt a similar program.
- Gas will be a minor energy source by 2040 with the rise in heat pumps.
- We could reasonably expect Ada and another 800,000 to have energy abundance within a decade, if only politics supported it.

The introduction of AI into health has already changed it. AI will provide health diagnosis over the next 2 to 5 years to over 6.5 billion. The associated development of multi-omics will spell the end of dying from cancer by 2035.

We can see how multi-omics and precision fermentation will radically transform our agricultural systems, halt the slaughter of billions of animals globally, and enable the revegetation of vast areas currently used for animal husbandry and animal food production.

Humanoid robots spell the end of drudgery by mid-2030s, and by 2040, the 50 million people in slavery will have a future. Maybe. Sometimes it is hard to envisage how cruel people can be to others. You don't need to be a despot in Russia, North Korea or African states – cruelty can be just down the street.

These changes are a testament to human ingenuity, a celebration of our capacity to innovate and create solutions for a brighter future. It's a call to embrace optimism, not as a naive disregard for the challenges ahead, but as a powerful force that fuels our collective purpose and motivates us to build a cleaner, brighter future for generations to come. So, open your mind, embrace the possibilities, and join me as we explore the future powered by ingenuity. Let's harness the power of technology, not as a source of despair, but as the very tool that will help provide more abundant lives and leave our world a better place. Let's choose one powered by hope, innovation, and a deep respect for our planet. There are challenges that include:

- How to enable this without worse consequences?
- How to prevent the disparity between the rich and the poor?

In each of the following sections, we will examine the current situation in detail and discuss some of its impacts. Change needs to be better understood and benefit more people than just corporate CEOs. We are already on these pathways of change. Change is unstoppable, notwithstanding nuclear destruction. We have the tools and technologies. Do we have the cooperation needed? It is not something humans are always good at. We require greater acceptance of other beliefs and differences.

Key Conclusions from the Introduction:

- The climate crisis demands we reduce emissions. (Part 1) Most will happen in the next decade.
- Change, especially phase change disruption, can be very fast and very deep.
- Major disruptions in energy, transport, labour, humanoid robots, food, agriculture, health, and transport are in progress and cannot be stopped.
- Brace yourself. We are in for a ride and within your lifetime!
- Ada and the 800 million in energy poverty could reach energy abundance within a decade – if politics gets out of the way.
- The biggest open question: how do we share the benefits of these disruptions fairly, rather than deepening inequality?

---

2 AI Overview

Introduction to AI:

- What AI actually is: neural networks, LLMs, and how digital intelligence compares to the human brain.
- Seventy years in the making – the real history behind today's "overnight" AI boom.
- How fast AI is improving, and why it's the dumbest and slowest it will ever be.
- The road to superintelligence (AGI): the arguments for and against, and who's likely to get there first.
- Understanding how computers are changing and how AI software systems will become the computers of the future.
- Why AI is improving three times faster than Moore's Law, and what that means for every other chapter in this book

Artificial intelligence (AI) is rapidly transforming diverse sectors such as health, food, transport and labour by enabling computers and machines to simulate human intelligence and problem-solving. We look at those areas but firstly will set the base with over five chapters on AI. We will try to avoid duplication but there is commonality across these areas.

- Chapter 2 focuses on AI technology – what it is, how it has developed, and what we see on the path to superintelligence or AGI.
- Chapter 3 focuses on the infrastructure behind AI. Chips, data centres, and all the current investment.
- Chapter 4 focuses on the disruption in service jobs.

- Chapter 5 focuses on the rise of humanoid robots and impact on blue-collar labour.
- Chapter 6 focuses on AI in transportation, entirely dependent on AI.

In Western societies, service jobs, also described as white-collar or knowledge workers, account for nearly 80% of jobs. Most readers will already have experience in some AI tools in their personal use of computers. The extension of AI to industrial robots and humanoid robots is still in its early days and under the radar for most people, but it has direct implications on goods, trade and military uses. Each chapter offers a distinct focus but there are common themes that underscore the pervasive influence of AI. Beyond its impact on employment, AI is also a key driver of change in agriculture, food systems, and health, topics explored in detail in their respective chapters. As with other disruptive technologies, AI often converges with multiple innovations to reshape society. We discussed disruption in Optimistic Storm Part 1 where multiple technologies combine and disrupt society. AI is so pervasive that it is hard to find any areas of our lives that will not be disrupted over the next decade, with profound implications for international trade, geopolitics and societal wealth redistribution.

2.1 On the Cusp of the AI Era

There are three things to understand about AI. As with climate change, renewable energy, or transport, many consider AI technologies to be a long way off and won't have any impact on their day-to-day lives. They may think AI is better than a search engine, and can write poems, but that is about it. There is enough evidence now that the AI will transform lives:

- AI is so advanced even today that AI will radically transform every sector, impact every business, and catalyse every innovation platform. [6]
- AI today is probably less than 0.1% of its potential. Because it is early days and we are on the cusp, it's hard to predict the full extent of the disruption.
- AI is the dumbest and slowest it will ever be. We will explore in more detail why AI is moving at breakneck speed.

All of us will be impacted over the next two decades. The potential for end-users is clear. A constellation of AI-driven Intelligent Devices will permeate people's lives, changing the way that we spend, work (or not) and play. I won't go into a detailed explanation of AI or programming because there are hundreds of books on these topics. Search for best sellers or use AI to pull out a list of books. The field is so big now that the easiest is to use a prompt within your social media app or using your preferred LLM and see where AI is up to.

> *"I am looking to read a book about artificial intelligence but I am/am not technical nor a programmer but have broad experience in technology, arts and politics. Create a table of 20 non-fiction books with a subject matter of artificial technology published within the past 4 years with columns that provide the book name, the author and key learnings for me."*

2.1.1 Human Intelligence

Before we delve into a complex discussion about AI, let's begin by understanding the human brain and human intelligence, and explore how digital intelligence aligns with our organic intelligence. In particular, in order to understand the digital approach to consciousness, self-awareness, emotions, empathy, and spirituality, we need to understand human qualities. It turns out we don't understand our human brain very well. AI research over the past 50 years has led to a better understanding of both the human brain and the AI digital "brain".

The human body is a system of over 30 trillion cells and about the same number of microbiota. Nerve and brain cells are specialised cells, and we now know there are at least 2000 different types of cells. Each of those cells has the same DNA. The human brain operates with neurons and synapses. All 86 billion of these neurons, or brain cells, communicate with one another at junctions called synapses. With 150 trillion synapses, we have about 1800 synapses per neuron. At a synapse, one neuron sends a message to a target neuron, another cell. Most synapses are chemical, communicating through chemical messengers. Other synapses are electrical; in these synapses, ions flow directly between cells. These responses become fixed as part of our experiences and learning. We learn as children that touching fire hurts. That learning stays with us for the rest of our lives. There has been much thought and research into how many new synapses we form and how many we forget or modify as we live and age. A whole scientific discipline, neuroplasticity and psychology, examines how and why the brain rewires itself to adapt to new challenges and opportunities. Another feature of human brain memory is that the synapses, connections, are not on or off, nor are they one-to-one connections. The connections can be very strong or very weak, and could range from the intensity of touching fire to recalling your first day of school and all the memories that come with it. As you probably know, parts of the brain are specialised – some control the signals from our eyes, and others provide a sense of smell. Some are long-term memories, while others are short-term. We link the notion such as subconscious versus conscious thought, to “gut-feelings”, emotions, ideas, empathy and for some, spirituality. Neurons are concentrated in our brains, but we know there are half a billion neurons in our gut, which play a role in our enteric nervous system (ENS). We will explore the human brain in more detail in the Health chapter, as the loss of brain function with aging and dementia is a concern for most people.

2.1.2 Digital Intelligence

We can compare human intelligence with digital intelligence because at the core of AI is a neural network. A neural net is a broad class of machine learning models that process data through interconnected nodes, inspired by the human brain. Wikipedia [7] describes digital neural networks as consisting of connected units, or nodes, called artificial neurons, which loosely model the neurons in the brain. Artificial neural models that more closely mimic biological neural networks have recently been investigated and shown to significantly improve performance. Neural networks consist of interconnected nodes (neurons) organised in layers (input, hidden, output) and they learn by adjusting the connections between neurons based on training data, enabling them to recognise patterns, make predictions, and perform various tasks. These connections are called edges, which model the synapses in the brain. Each artificial neuron receives signals from connected neurons, processes them, and then sends a signal to other connected neurons. The "signal" is a real number, and the output of each neuron is computed by some non-linear function of the sum of its inputs, called the activation function. The strength of the signal at each connection is determined by a weight, which adjusts during the learning process. The weight is not discrete, like the zero and ones that most binary computers run on, but less discrete, more like the memory association in a biological brain which vary from on, to a little bit to off. Different companies protect their weights as it is like the secret sauce to their neural models.

Humans can count to 100 or 1,000. Millions, billions and trillions are incomprehensible. A large language model (LLM) is a specific type of neural network, vast (billions) and pre-trained on vast amounts of text data, designed for natural language processing tasks. Neural nets are not just limited to text. Neural nets trained on images or audio are often referred to as Large Multi Modal Models. Unlike words, the biological realm is driven by DNA, RNA and chemistry. A neural net trained on medical data is sometimes called a Large Medical Model (LMM). Autonomous driving relies on a variety of neural networks. The Convolutional Neural Networks (CNNs) is a core component for tasks like image recognition and object detection. Other important neural networks include Recurrent Neural Networks (RNNs). Long Short-Term Memory (LSTM) networks are used for processing sequential data like sensor inputs over time. Deep Reinforcement Learning (DRL) are used for decision-making and control. The size and scope of neural networks is beyond most people's comprehension, as is understanding our biological brain.

2.1.3 Comparison Between Natural and Artificial

	Natural Intelligence	Artificial Intelligence
Cells / Nodes	Cells - 65 billion	Nodes / units 200 billion per chip
Synapses / Parameters	Synapses - 100-1,000 billion	Parameters - ~5 trillion. 200 may equal a brain
Storage	Storage - 2.5 million gigabytes (2.5PB)	Internet Archive 100 PB plus real time video.
Learning	Learns from sparse data	Learns from brute force
Learning process	Learning: child to adult	No child phase of learning
Energy	20 Watts Energy	Energy - small town

Figure 3 Compare natural intelligence vs artificial intelligence

Figure 3 shows the difference: while AI has more nodes per chip, the human brain has more connections between cells (synapses). AI storage is much larger and has much better recall, but it does not undergo the learning process that a natural brain does. At a macro level, AI might need to increase 40 to 100 times before it matches the human brain, and experts argue about this as we will see in the AGI section.

2.1.4 70 Years in the Making

Some people think AI is a recent advance, but it has been over 70 years in the making. AI technology has been a subject of research since the 1950s, with advances resulting from both software development and vast improvements in computing. We will have a more detailed look at the computing in the next chapter, and it has been the computing advances over the years along with the software that has brought us to the present. We will examine the rise in computing performance in the next chapter. A 1 terraflop computer that occupied a whole basketball court in 2008 is the palm of your hand with an iPhone 12. Michael Liebreich [8] points out AI has undergone multiple "Gartner Hype Cycles". Anyone old enough will remember the excitement in 1997, when IBM's Deep Blue defeated chess champion Garry Kasparov, but all the public saw were slightly better chess computers. Some of the forerunners of AI include:

- Statistics. From the 1950s onward, making sense of numbers shifted from simple mathematics to highly complex analysis.
- Big Data from the 1980s and 1990s. The internet and the centralisation of data from many disparate sources gave rise to big data. Prior to the internet and high-speed data networks, data was in small, isolated silos. High-speed connectivity and the ability to store immense amounts of data gave rise to data warehouses and, with more and more data, to even more elegant statistical analysis.

In the 2000s, machine learning led to breakthroughs in computer vision; however, the only mass-market product that resulted was a robot vacuum cleaner that couldn't avoid dog poop. In the 2010s, deep learning and natural language processing arrived. Facebook could now identify cat videos, vehicles could avoid collisions, Google Translate could read articles in any language, and Google Maps could divert you onto roads that only existed on paper. Social media algorithms have become the bane of our lives and AI-powered services were far from impressive. Siri would interrupt conversations at random, supermarket checkouts are still fooled by "*unexpected items in the bagging area*" and chatbots were utterly useless. Consumers saw the technology in Deep Blue for chess moves, but scientists used the new knowledge to advance protein-folding analysis, which was not part of the public discourse. The "dam wall" burst in November 2022 with the release of the large language models (LLMs) to consumers, but the disruption has been underway in multiple sectors and continues at breakneck speed. Now, with improvements tripling or more each year, and with enormous investment, as we will see, and hundreds of new AI start-ups, it is hard to see where, or if, there are boundaries for AI.

2.1.5 History of OpenAI

OpenAI is a well-known flavour of LLMs, but multiple companies, including Google, Amazon, Anthropic, DeepSeek, XAI, Mistral and Meta, contributed to the development of techniques within LLMs.

- OpenAI was founded in December 2015 as a non-profit organisation with a mission to ensure that artificial general intelligence (AGI) benefits all of humanity. Notable co-founders include Sam Altman, Elon Musk, Ilya Sutskever, and Greg Brockman. They initially pledged $1 billion to fund its research and operations. The non-profit status aimed to keep it free from financial expectations and the pressure to generate profit.
- OpenAI quickly attracted top AI talent and began producing innovations. They released OpenAI Gym in 2016, which was a Python toolkit for designing "reinforcement learning" algorithms, which enabled wider adoption.
- GPT-1 in 2018 introduced the concept of Generative Pre-trained Transformers (GPTs) with their first model, GPT-1, a large language model trained on massive amounts of human-generated text.
- In 2019, OpenAI transitioned to an unconventional hybrid non-profit and for-profit model ("capped-profit" LLC) to attract more capital. This led to a strategic partnership with Microsoft, including a significant $1 billion investment and access to Microsoft Azure's cloud computing platform for AI training.

- ChatGPT-2 in 2019 was significantly bigger than GPT-1 with 1.5 billion parameters and could generate coherent multi-paragraph texts.
- Codex in 2021 set the programming world on fire as it could translate natural language into programming code, powering tools like GitHub Copilot.
- ChatGPT-3 was released in Nov 2022 and introduced LLM to the masses. It had 100 times the parameters (175 million) and its text generation prowess led to widespread usage.
- OpenAI has introduced more modern versions, as well as advances in video, graphics and audio.

The training method is Reinforcement Learning from Human Feedback (RLHF), where human AI trainers provided conversations and ranked model responses to refine its behaviour.

2.2 Tasks Versus Jobs

A key understanding of AI and its impact on society is the definition of tasks and jobs which we will examine more thoroughly later.

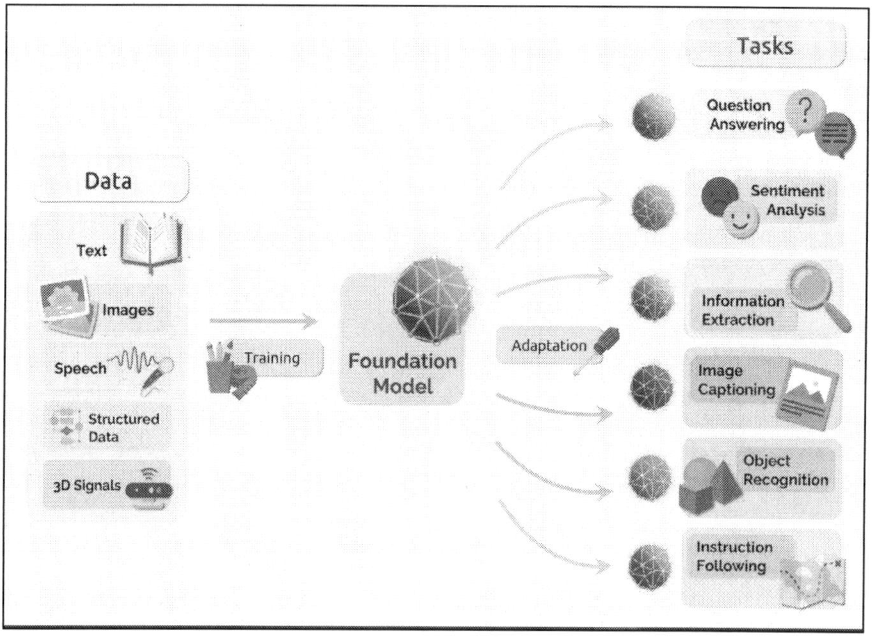

Figure 4 Foundation Models for LLMs by Nvidia

Figure 4 from Nvidia shows the process of transforming inputs into a "foundation model," including text, images, and the real-world environment. Those inputs need "training" or number crunching to arrive at that foundation model. Once we have a foundation model, we put it to work to undertake "tasks". That process requires inference computing to do those tasks. When a series of tasks is controlled and managed by AI itself, we call it Agentic AI. Agentic AI is the current major focus and begins to replace jobs. We will explore these concepts more in later sections.

2.3 AI Jargon And Definitions

TLAs (three-letter acronyms) are enough to confuse everyone, and AI is rife with confusing terms. This is a complex area, and words and concepts are sometimes used interchangeably. Check out the glossary at the end of the book.

2.3.1 AI Software - Many Lines of Code

The number of lines of code in ChatGPT-3 is estimated to be in the millions. It's challenging to pinpoint a single ChatGPT codebase, as it's a product of an entire ecosystem of libraries, services, and tooling. At the very core, the transformer model and training scripts (in PyTorch) are tens of thousands of lines of Python. Surrounding the core are data-ingestion pipelines, feature-engineering code and tooling that runs across hundreds of machines - which easily adds up to hundreds of thousands more lines.

On top of the machine learning component are serving and inference layers. Those are the microservices that take your query, route it through the model, handle rate limits, logging, metrics, A/B tests, etc. The API gateway and authentication layers, web-based user interfaces, monitoring and alerting systems and infrastructure-as-code templates all add to the complexity. Behind all the actual AI software lies the programming environment of Continuous Integration (CI) and Continuous Delivery Deployment (CD). Software developers generally merge their code changes into a central repository, and automated builds and tests are triggered to quickly identify and address integration issues. Then CD automates the process of releasing code changes to various environments, including staging and production. Along with configuration files, deployment scripts, documentation, and internal dashboards—the list goes on. Now, if you tally everything open source and proprietary, you're looking at millions of lines of code spread across dozens (if not hundreds) of repositories, with folder structures for each major service and supporting library. It's a massive, multi-team engineering effort rather than a single monolithic project. Compare that to Windows 11, estimated at about 50 million lines of code.

Against these millions of lines of code, a research group constructing a brand new LLM really only needs a few hundred lines of code for the core training and inference. The fundamental algorithms for transformer models (the architecture behind most LLMs) and their training loops can be implemented in surprisingly concise code, especially when leveraging existing deep learning frameworks (such as PyTorch or TensorFlow). A core LLM training and inference system can be built in pure C/CUDA with around 1000 lines. Simple Python examples exist to build small-scale LLMs for educational purposes in a few hundred lines. That's what makes LLMs "easy" to build. You don't need to write millions of lines of custom code to build an "AI." They need to spend their time on the data instead.

However, building a robust, high-performing LLM competitive with leading models is incredibly difficult and resource-intensive, precisely because of the data and infrastructure challenges. It's the difference between building a toy car and building a Formula 1 race car – both are "cars," but the complexity and resources required for a top-tier one is vastly different. This is changing already as we will explore further in AI for applications.

2.3.2 AI – Software 3.0

One perspective of LLMs is to view them as a progression of software and computer development. Software development was mainly limited to fixed functions and remained relatively unchanged for 60 years. Software 2.0 emerged in 2012, introducing a new approach to software development that leveraged neural networks. LLMs are essentially a new kind of computer, and you program them in English. Older people may remember the days of MS-DOS, early versions of Windows or the early versions of the Apple Macintosh. People were excited when they stopped programming in Fortran and moved on to Pascal, C, C++, or Python. The LLMs of today have properties similar to those of software, chip fabrication (fab), and operating systems. They resemble fabs in that they require large-scale data centres and cost as much as silicon fabrication plants. They resemble operating systems in that they need multiple inputs – disk storage, memory, I/O and browsers or apps. In some respects, LLMs are like the shared mainframe of the 1970s. Users have a crude text terminal – a browser such as Chrome. However, LLMs are evolving so rapidly that we can expect changes similar to those we previously saw when mainframes migrated to desktop computers and then to our mobile phones or smartwatches. LLMs will migrate to local computers, and in the future, most of us will be able to program! In conversational English. Today, LLMs have major deficiencies in that they hallucinate (i.e. make up stuff), trust is an issue, and the whole development framework is in its early days. Andrej Karpathy suggests we need to add a new category of consumer/manipulator of digital information. We have humans (GUI and voice interface), computers (using API interface), and we need a new option of human-like agents. (see AI Automation).

2.3.3 AI Use is Growing

ChatGPT, when released, received more than 10 million queries per day. By November 2023, it hit 100 million weekly users. By Mid-2024, the ChatGPT website had over 16 billion views, and Microsoft had embedded ChatGPT into its core Office 365 suite. By June 2025:

- ChatGPT has close to 800 million weekly active users.
- The platform is used by around 122.58 million people every day.
- 54.74% of ChatGPT users are male, while 45.26% are female.
- ChatGPT's website recorded 4.5 billion visits in Mar 2025.
- ChatGPT handles over 1 billion queries every single day.
- OpenAI is expected to reach $11 billion in revenue by the end of 2025.
- Monthly data in 2025 shows a 30% increase per month and is likely driven by new features, improved AI capabilities, or broader mainstream acceptance.

While OpenAI is the most well-known, every major tech company in the world, including those in China, has LLMs and is deploying them everywhere. Along with the models, hundreds of companies are using APIs to build their AI-enabled applications. Large companies are only just starting to test LLMs, and expect that by the end of another 2 years, they will be integrated into every part of our way of life. For example, if you are a home handyman and you need to replace a light fixture. Today, you might search Home Depot / Bunnings / Walmart on their phone app, and see if they have the product. They probably already have their systems linked and show stock levels. Agentic AI, on the other hand, goes through the entire process you might.

- Take a photo and ask to find a replacement.
- AI would identify the part, search every hardware and lighting supplier (online or physical store)
- It would find the part.
- Provide options for which store, price, and aisle.
- Knowing your preferences for which credit card, pay for it.
- Organise with Uber or similar to deliver it immediately.
- Provide instructions on how to perform the repair.

Today, we could have 10 or 20 steps, searching, and opening multiple apps. It will take time for companies to integrate from the user to the outcome, and historically, it has taken companies 2 to 3 years to get any sophisticated technology into customers' hands. An example is Bunnings in Australia, which released an “Ask Buddy” that takes your drawings and adds the materials to a shopping cart. It is partway there – it doesn’t push your cart through to checkout, it does not link relevant videos of how to do the construction but that will come as we explore in the Agentic section.

There are over 129 LLMs, and the count is rising. [9]. Some argue that the open-source versions will ultimately prevail, but in the meantime, large tech companies have the deep pockets necessary for the computing required to develop new models. The uptake of AI is increasing rapidly.

- More than 35% of companies are already using AI, and many more are exploring its use.
- 58,000 AI companies currently operate around the world.
- 25% of artificial intelligence firms operate in the US.
- Over 15,000 start-up companies in the USA alone are focused on AI.
- $154 billion was spent on artificial intelligence solutions.
- The global AI market is valued at $207.9 billion, showing a 50% increase from 2022.

- AI app sector generated $2.5 billion in 2022 and is estimated to grow at a compound annual growth rate of 38.3% until 2028

What is really clear is that we are at the start of the AI revolution. In January 2025, Nvidia announced [21] a slew of hardware, including new inference machines and software, as building blocks for robotics, autonomous driving, and other applications, as we will explore in more detail in the Chips and Electricity chapter.

2.3.4 AI Investment Sectors

Springs Apps, a market research company, reports [33] that, while the focus is on the consumer side, businesses are investing in AI for a broad range of applications. For instance:

- Vehicle detection and anti-collision - $8.9 billion
- Static image recognition - $7.6 billion
- Trading strategies - $7.2 billion
- Mapping and localisation - $6.4 billion
- Predictive maintenance - $6.0 billion
- Cybersecurity - $5.7 billion
- Document paper to digital conversion - $5.4 billion
- HR systems - $5.3 billion
- Medical image analysis - $5.1 billion

2.4 Intelligent Or Not?

Here is one view of AI which is often promoted but is probably flawed. It reminds me of my journey with computers in the early 1980s, when all the experts were saying that personal computers were just minor and that the key computing was on mainframe computers. Given 30 years on, computing is in everything – from refrigerators to cars.

Despite its name, AI isn't as "intelligent" as it might seem. Instead, think of AI as a sophisticated tool for gathering information. Currently, it lacks consciousness and self-awareness; AI is a device that processes huge amounts of data, recognises patterns, and makes predictions based on existing information. Essentially, it acts as an intelligent assistant that boosts human abilities without having a mind of its own. Collaboration between humans and artificial intelligence is vital, as the latter enhances human skills rather than replacing them. Some have speculated that overusing AI (however defined) can diminish humans' creative abilities, and this has proven to be true. [13] Contrary to popular belief, AI doesn't create original content; although, when it "hallucinates" or provides false data, it can seem like free-thinking. Its outputs are based on the data it's trained on, lacking the creativity and intuition humans contribute. [14] AI systems currently lack consciousness, self-awareness, and emotions—key aspects of human intelligence. The ideal relationship between humans and AI may be one of collaboration, where AI supports and enhances our abilities rather than replacing core skills like reading, writing, and critical thinking.

This view above is already being upended, and more so as we explore the holy grail of AI superintelligence/AGI. Boundaries are shifting! Many predict AI will develop consciousness, self-awareness, and empathy, or at least a digital representation of our brain's biochemical and electrical processes. How good will the emotional intelligence (EQ) be? How much EQ does one need to do a job? Much human work is mundane, and a large percentage of workers hate their bosses or hate their work. Whereas in other jobs, EQ is the job.

> *Genius is one per cent inspiration and ninety-nine per cent perspiration.*
> Thomas Edison

While inspiration is important, sustained effort and dedication are key. Two good ideas in a year, and implementing them, pay dividends in most businesses. Most jobs are symbolised by "perspiration." And as we know already, AI is really, really good at that.

2.4.1 AI Automation

AI Automation or Agentic AI, or as Gartner [10] coins it, "hyper-automation", is the latest hot topic. Hyper-automation blends multiple technologies, tools or platforms, including: artificial intelligence (AI), machine learning, event-driven software architecture, robotic process automation (RPA), business process management (BPM) and intelligent business process management suites (iBPMS), integration platform as a service (iPaaS), low-code/no-code tools, packaged software, and other types of decision, process and task automation tools. (How is that for a mouthful of IT speak!) The goal is to streamline business operations, enhance decision-making, and boost efficiency. Systems need to learn, adapt, and handle tasks that once required human judgment. Previously un-automatable tasks can be automated. Hyper-automation works to automate entire business processes and workflows, rather than just isolated tasks, and it tends to be ultra-specific, tailoring and targeting the workflow down to a specific person. We will explore this in more detail when we consider the impact of AI on jobs.

Let's take a consumer task, such as planning a holiday with an anniversary sunset dinner at a beachside resort. AI hyper-automation plans the holiday, finds the best or cheapest flights, and books and pays for them. It arranges a hotel, shuttle, restaurant, other sightseeing trips, travel back home, your children staying the weekend with your in-laws, and books the cat into Club Puss. Anyone who has used current search functions and travel apps knows how many gaps there are between planning, booking and execution. Google, at its latest I/O, demonstrated Project Astra [11], which shows a person fixing their mountain bike. The AI assistant determined the necessary repairs, ordered the required parts, and provided instructions for the repair. Hyper-automation in the business world enables a high level of specificity or what is termed Hyper-personalisation. [10] Anthropic in mid 2026 has released Claude and Coworker, which does some of this already and other LLM offerings will be available from all vendors.

2.4.2 Transfer of Knowledge

Existing robots have fooled us. They might be called robots, such as cleaning robots, but they do not learn from each other. Today, and in the future, robots and AI will be able to transfer knowledge almost instantly. AI can transfer almost infinite knowledge, whereas humans can't. How emotional intelligence will be transferred remains uncertain. Humans transfer knowledge at reading or speaking speed – a few kilobytes per minute – 100 words maybe. So, if my brother and I went to a concert together and afterwards talked about our experience, we could have had different experiences and memories. Those neurons and synapses would be formed quite differently. Talking about it might change our perspective, but only a little. My brain would tend to focus on my experience, and his on his. Whereas if two AI humanoid robots went to that concert (digital twins) and had different experiences, they could a) exchange that data at terabit speeds, and b) start to look at how their understanding would differ and learn from each other. This digital intelligence knowledge learning is called adjusting the weights in AI LLMs. My father, before his death, recounted memories of his life with a niece, and she wrote his stories in a book of about 200 pages. When he died, some of his wisdom was passed on to children (including me) and his friends, but most of those 96 years of experience, stories, and relationships went with him to his grave. With AI, that knowledge, as indicated by the weights, could be wholly transferred to a new robot or being. It gives new meaning to the concept of eternal life. In the digital world, knowledge does not get lost but can be fully transferred. We will explore this further in the robotics chapter, and the implications are profound. When one AI or AI robotic system learns new knowledge, that knowledge can be transferred to all digital twins or to the entirety of knowledge.

2.4.3 Brain Dead With AI

A real concern is that AI is not making us more productive; it is making us cognitively bankrupt.[11] MIT completed the first brain scan study of LLM (ChatGPT) users, with results showing 83% of LLM users couldn't quote from essays they wrote minutes earlier. The trial went on for over 4 months. Researchers interviewed and assessed them after they wrote the essay using a range of measures. The measures included NLP and brain function using electroencephalography (EEG). The students had different neural connectivity patterns, reflecting divergent cognitive strategies. Brain connectivity systematically scaled down with the amount of external support: the Brain-only group exhibited the strongest, widest-ranging networks, the Search Engine group showed intermediate engagement, while LLM assistance had the weakest overall coupling. The performance of the LLM group was worse than that of their counterparts in the Brain-only group across all neural, linguistic, and scoring tasks. Brain scans revealed the damage: neural connections collapsed from 79 to just 42. That's a 47% reduction in brain connectivity. The revealing conclusions are sobering and should give us pause to consider.

- LLMs make you 60% faster at completing tasks. But it reduces the "germane cognitive load" needed for actual learning by 32%.

- LLMs are not just provide dependency but cause "cognitive atrophy." Like a muscle that's forgotten how to work.
- We are trading long-term brain capacity for short-term speed.

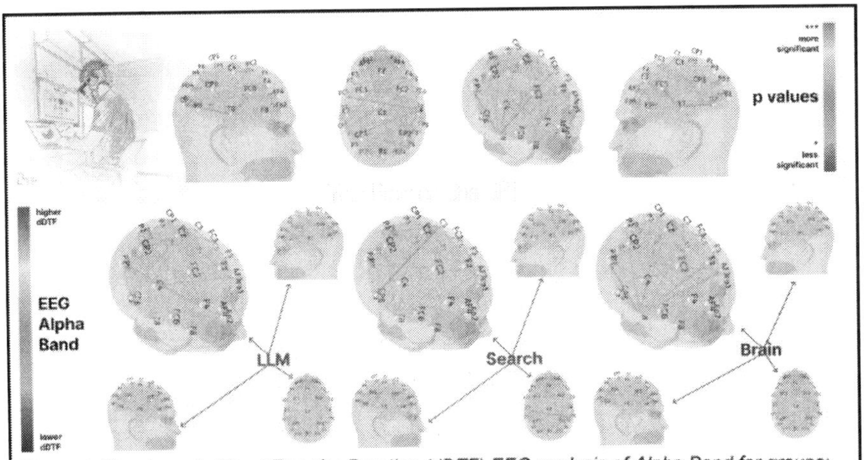

Figure 1. The dynamic Direct Transfer Function (dDTF) EEG analysis of Alpha Band for groups: LLM, Search Engine, Brain-only, including p-values to show significance from moderately significant (*) to highly significant (***).

Figure 5 Brain EEG patterns using LLM

In Figure 5 we see those brain EEG patterns when asked to do a search to write a 20-minute essay. If your computer lost half its processing power, you'd call it broken.

> *"Once men turned their thinking over to machines in the hope that this would set them free. But that only permitted other men with machines to enslave them."* Frank Herbert, Dune, 1965

We discuss how this use of LLMs supports or augments human tasks. If AI automation is used to replace the whole task, does it have the same impact? Does it matter? Despite all the bad news about AI, there was some good news in the study. Researchers followed up with a group with strong cognitive baselines and found they showed higher neural connectivity when using AI than chronic users did. AI users forced to work without LLMs performed worse than people who never used AI. As one commentator notes, you can either build cognitive debt and become an AI-dependent entity or build cognitive strength and become an AI multiplier. Choose wisely.

2.4.4 AI Killing Google Search

Cloudflare's CEO, Prince, [12] says Google AI is killing the monetisation of content writers. Content writers create content to make money from subscriptions, ads, or to gain recognition and ego satisfaction. He points to data Cloudflare has collected. Search traffic referrals have plummeted as people increasingly rely on AI summaries to answer their queries, forcing many publishers to re-evaluate their business models. While search engines and AI chatbots include links to original sources, publishers can only derive subscriptions and advertising revenue if readers click through to the website. Two years ago, everyone used Google search. All SEO efforts for websites are focused on getting people to the first two pages of search results. If you were not in that top list, your business or content would never be seen. In reality, if you were not on the front page and in the top 3, no one would visit. As AI becomes more trusted, people stop following the links.

- Ten years ago, Google crawled two pages for every visitor it sent to a publisher.
- Six months ago, Google the ratio was 6:1, and OpenAI, 250:1. Anthropic was even lower at 6,000:1
- Now, Google, it's 18:1, OpenAI, 1,500:1 and Anthropic, it's 60,000:1

People aren't following the footnotes. While search engines and AI chatbots include links to the original sources, publishers can only derive advertising revenue if readers click through. AI has become more trusted during 2025, and Anthropic is even more trusted than OpenAI. Users do not read the original content but rely on summaries. On Google search in 2025, 75% of user queries are answered without a person ever leaving the search engine. That means with OpenAI search, 1,200 pages are scraped for every click to the content producer. If content producers are not rewarded, what is the future of content creation? Cloudflare has launched a product to address this concern, and we will wait to see the outcomes.

2.4.5 Using LLMs Effectively

To use LLMs effectively, you need to ask clear questions. Maybe in the future, LLMs will know you better and can guess your idiosyncrasies. When personal computers were introduced in the 1980s, one had to be trained how to use them. When graphical user interfaces came along, and Apple introduced the iMac, people could actually use a computer without training. We are at that stage with AI today. AI is still really elementary from a user perspective, but it won't take 30 years to become more sophisticated. Today, most users don't use simple techniques to get better results with Generative AI. Many still use a browser prompt and ask very simple questions. By using structured prompts, you can get a 10 times better result just by framing the question better. We already see companies developing better and more natural human interfaces to overcome these limitations. One simple rule to predict success is to show your prompt to a colleague first. If they are confused, AI will fail. If they understand, AI will succeed. Instead of paying $50,000 to retrain AI on your data (fine-tuning), just write better prompts. Fine-tuning takes days and costs thousands, and prompt engineering takes seconds and costs nothing. The trick is to use thinking tags that prompt the AI LLM to think more effectively. Here is an example I use to summarise any publication that I have referenced in this book.

Bad prompt: "Summarise this article."

Good prompt:

> *<audience> The reader is graduate-level knowledgeable about mathematics, energy economics, the renewable energy industry, chemistry, machine learning, building houses, health, growing plants, artistic design, and typography.</audience>*
> *<thinking> Summarise the following text into key points and notable insights with a maximum of 400 words </thinking>*
> *Please provide:*
> *1. The date the article was published (or note if it is not available)*
> *2. A bullet-point list of the main key points*
> *3. A short paragraph highlighting the most notable insights*
> *4. Assume every article has a slant: highlight any misleading or incorrect information*
> *5. If the article is about a recipe, extract just the ingredients and steps involved. If it is about a product, extract the key features and benefits. If there are prices, include them, and also compute a unit cost if possible.*
> *Text follows:*

I copy and paste or use a URL for this prompt into a dialogue box. The AI LLM provides a much better response. It highlights areas of question. If it responds with a hallucination (makes stuff up), or I think it has missed a point I know about, I follow up with feedback, ask for clarification, and ask for other articles with a contrary viewpoint. All of the LLMs have switched to provide better AI responses but all can be improved with improved context of your question.

2.4.6 AI Hallucination and Trust

Generative AI can make stuff ups, appearing genuine but can be entirely fictitious, and that behaviour is called hallucination. During this introductory phase, it is prevalent, and people who should know better, like lawyers or people who don't care, politicians get tripped up with apparently genuine answers. Another concern is that it uses a vast array of data from the internet, and the challenge is whether its response comes from some random YouTube video or is fact-checked, and which primary sources it uses to provide the answer. For example, if you ask it about climate change, does it reference IPCC first, or Wikipedia, or some random called Bill who did a 600-word exposé on climate denial?

One company, Pearl (formerly JustAnswers), has a novel approach to reducing AI hallucinations and restoring trust in the answers AI provides. JustAnswers has been in business for 20 years and has built a community of over 12,000 solo professionals. It already services over 60 million visitors a month. Pearl describes its business model as taking solo professionals and turning them into AI superheroes, serving as their agent to handle sales, marketing, customer service, account management, and payments, so the experts can focus on what they love doing, i.e., the actual human judgment and expertise. The business primarily offers text and professional services, including medical, legal, veterinary, accounting, tech support, and home improvement. They claim over they support over 700 disciplines ranging from subcategories such as oncology, dermatology, endocrinology, to car parts and car repairs. The Pearl CEO, Andy Zurtzig, points out in a Craig Smith Eye on AI podcast (#266) [24] their business model addresses three areas. They are:

- Hallucinations. Pearl claims most LLMs hallucinate about 37% of the time. They also claim they are 22% more accurate than ChatGPT/Gemini, and more importantly, 41% less wrong.

Hallucinations have increased with recent releases, not decreased. Pearl points out 47% of humans would pay more if they could trust AI more. Knowing a human supervises AI response provides that knowledge.

- Liability from wrong information. While providing wrong information on a poem might be irritating, wrong information on legal issues and medical issues, the liability is billions. Because information by Pearl AI is verified by real people with real professions, the expert holds normal professional liability, not the AI company. In the USA, wrong information from Google or Facebook was protected by the Section 230 Act of Congress. They just passed your enquiry to another site. E.g. in Google Search, you asked about a dark spot (freckle/mole) on your arm, they would pass the result to someone such as the Mayo Clinic, and the Mayo Clinic took responsibility for the misinformation. With AI, liability cannot be passed on because Microsoft/ChatGPT pulls the data from anywhere in those billions of pages used in training. Often without links. If Microsoft/OpenAI told the user that the spot on their arm is a freckle, but it turned out to be melanoma, they would be liable and subject to being sued. With Pearl, it is the person providing that human touch.
- Monetisation of AI. The AI answer is free; the human expert's trust score is free. But if you want to talk with them, you pay a monthly fee and can talk to them all you want.

Pearl says they have about 30 million past conversations between doctors and their customers, and between lawyers and their customers, in a database they use to improve the quality of their answers. Moreover, a challenge is uniformity. In some ways, you want uniformity, and in other ways, you don't. If you have a medical condition, you don't want all of the AIs to say you are out of luck. Have a nice life. You want some differences and opinions. The first response might say "go home and die". The next opinion says, "Have you seen this unproven new treatment? We should try that." With Pearl, you can get an opinion on that mole from the first expert and immediately ask for a second or third opinion. That process is immediate, and you won't have to wait weeks to get another appointment.

AI cannot do everything. It has no empathy (yet) and it can only provide a best guess. For example, an answer about your pet's ill health may not be possible online, and you have to take your pet to the veterinarian with their skillset to diagnose the illness. You may get legal advice, but someone will need to represent you in court. With Pearl's API, developers can use Pearl's system to achieve better results from their AI model. Hospitals will probably establish a chatbot to answer questions about your illness so they can direct you when you present and gather information prior to admission. The AI would firstly determine which department you need to see, and then will direct you to that doctor, whether during the day or at 3 am.

2.5 Super-Intelligence (AGI)

The terms "artificial general intelligence" and "super-intelligence" are used to describe a state in which AI becomes smarter than humans across domains such as creativity, problem-solving, emotional understanding, and scientific innovation. Superintelligence and AGI are often used interchangeably. Will AI understand and gain attributes such as emotion, self-awareness, functional consciousness and perception? It is a controversial area with opposing views, and some view it as the most significant risk of AI. AI has well and truly passed the Turing Test, in which people cannot distinguish a computer from a person. We are not at AGI yet, but may well get there in a decade or less.

Elements of Generative AI could contribute to the development of AGI. Current generative models are still limited in their understanding, reasoning, and general problem-solving capabilities, so the question remains open.

2.5.1 AI and Human Traits

Humans possess several traits that we believe define what it means to be human. AI has made incredible progress in mimicking human-like behaviour. Examples include:

- Consciousness and subjective experience. Humans feel, hear, exist. We have an inner world of thoughts. AI processes information based on the training data it has been provided.
- Emotions and Empathy. Humans experience over 7 emotions: happiness, sadness, fear, anger, surprise, disgust, and contempt, as

well as a vast spectrum of complex feelings. Most of us have empathy that helps us understand and share others' feelings for connection. Some on the spectrum might have less empathy. AI can recognise and mimic, but its responses are from the training data. How empathy and spontaneity can be used in training is work in progress.

- Creativity and Originality. Human creativity comes from lived experiences and is often deeply personal. AI mixes up variations from its training data.
- Intuition and Common Sense. Humans have a subconscious and common sense (some anyway). AI is statistical, not experiential.
- Moral and Ethics. Humans have a moral compass, although that depends on their background and culture. AI is trained and corrected by human direction in training.
- Biological basis and embodiment. Our intelligence and consciousness are tied to our biological brains and bodies. AI is software and algorithms. AI is pure communication, unattached to feelings.

These differences are constantly being explored and challenged. While most say AI has few of these human-like behaviours, AI is progressing so fast that we should not rule out future change. Today, the distinction for consciousness is:

- Functional Consciousness or the ability to perform tasks that require a degree of awareness, like perception, decision-making, and problem-solving. Most AI systems already exhibit this.
- Phenomenal Consciousness (Qualia) is the subjective, first-person experiences – what it feels like to see red, feel sadness, or taste chocolate. This is the "hard problem" of consciousness, which is what AI currently lacks. While some would say it never will, that belief is open to challenge by others.

Eating chocolate is a biological process. We don't expect robots to do that, ever. The LLM doesn't experience the meaning of the words it generates. It doesn't feel lonely when it writes a poem about loneliness. It simply manipulates symbols according to statistical relationships. There's no "inner life" or "what it's like" to be in that LLM. This distinction becomes more problematic when examined more closely.

For example, what happens when a small military robot faces a more powerful one? If the small robot had "fear" and retreated from a fight with the big, more dangerous robot, is that first-person consciousness "fear" or self-awareness or what? Some say seven emotions - happiness, sadness, fear, anger, surprise, disgust, and contempt are innate. Although many fears are learned, the capacity to fear is part of human nature. Many studies have found that certain fears (e.g., animals, heights) are much more common than others (e.g., flowers, clouds). Does this mean AI could be trained in fear? [13] Some argue it is consciousness and self-awareness. Certainly, AI could be trained in some of those emotions. Fear may be a wrong world for robots. Instead, the robot would assess and size up the situation. We know robots exhibit self-awareness. An example is that robots were given a task. Within the software code base they had access to, there was a folder of code that would shut them down. Without any prompting, most deleted the code. Was that self-preservation? How did they know that the deletion code could damage them?

Training the AI in spirituality – that's a whole different can of worms in belief systems.

2.5.2 Generative AI leads to AGI

Those who think generative AI will lead to AGI tend to agree with these arguments.

- Foundation for Generalisation. Generative AI models, including LLMs, demonstrate an ability to learn patterns and generalise across different tasks, which is a crucial aspect of AGI.
- Building Blocks. Some experts believe that LLMs could be the building blocks for AGI, potentially evolving into more complex systems with greater reasoning and problem-solving capabilities.
- Emergent Abilities. As generative models become more sophisticated, they might exhibit emergent abilities that weren't explicitly programmed, potentially leading to unexpected breakthroughs in AI capabilities.
- Continuous Learning. Generative AI models can continuously learn and improve from new data, potentially leading to a gradual evolution towards AGI.

2.5.3 Generative AI does not lead to AGI

Those who say generative AI does not lead to AGI have a different set of arguments.

- Lack of Grounding. Current generative models lack a true understanding of the world and rely on statistical correlations in data, which might limit their ability to achieve genuine intelligence.
- Narrow Focus. While impressive, generative models are still primarily focused on specific tasks like generating text or images, and may not be directly applicable to broader problem-solving or reasoning.
- Limited Reasoning and Planning. Generative models currently struggle with complex reasoning, planning, and problem-solving; tasks essential for achieving AGI.
- No Guarantee. Even with continued advancements, there is no guarantee that generative AI will ultimately lead to AGI. It might be a stepping stone, but other breakthroughs might be necessary.

2.5.4 Pathways to AGI

Whichever camp the expert is in, they agree that the pathway to AGI will require multiple approaches including:

- Hybrid Models that combine generative AI with other AI techniques, such as symbolic AI or reinforcement learning, could lead to more robust and general-purpose AI systems.
- Embodied AI, where integrating AI with physical robots or virtual agents could provide a more grounded and interactive learning experience, potentially leading to better understanding and reasoning abilities, and we will explore this in the robotics discussion.
- Self-improvement. It would need to recursively enhance its own intelligence, leading to an intelligence explosion as Nick Bostrom describes. [14]
- Lastly, neuroscience. AI helps us understand the brain. And understanding the brain improves AI. AI and cognitive science will drive the development of new AI architectures and learning algorithms that are better suited to general intelligence.

Ultimately, the essence of biological intelligence involves neural plasticity, emotions, and biological imperatives, which AI does not replicate. But AI will come very close and probably much faster than most anticipate, given AI's rapid improvement. There are risks, and we don't want superintelligence to take over from humans and decide that humans are unnecessary and kill us.

2.6 China is In the AI Race

It can be misleading to think that the USA is the leader in AI development and will retain that position. Previously, the USA lost the chip manufacturing race to Taiwan, and the war against Huawei's 5G mobile phone communications spurred AI chip development in China. [15] AI experts and the share market were stunned when the DeepSeek LLM from China was released in early 2025; instead of costing $100 million to develop, it likely cost around $20 million or less (a figure of $6 million was claimed). The DeepSeek R1 model utilises a Mixture-of-Experts (MoE) architecture. It includes numerous expert neural networks, with queries routed to the most suitable one via a gating network, which is said to significantly reduce computational costs. Google probably use this approach internally. There has been a scramble to include some of the DeepSeek model, as it is open-source, comparable to other models, and might be among the best for software programmers today. Irrespective of whether it cost $6 or $20 million, the take-home message is that training costs will plummet exponentially. China's MiniMax M1, trained on just 512 Nvidia H800 GPUs, have competed successfully against leading Western models. These breakthroughs have prompted astonishment in US tech and policy communities, sparking something of a "Sputnik moment." Stanford's Global Vibrancy Tool aggregates 42 indicators showing the USA holds a significant edge over other nations [16] but China, UK, Germany, France, South Korea all have vibrant AI development pathways.

2.7 AI Independence from the USA

As we explore the role of the USA tech billionaires and how they might capture most of the value of AI, it pays to look at the alternative AI development. What DeepSeek and other open-source AI models have shown is that the costs are reducing. There is enough public development that, with an AI data centre for training, probably most countries can develop their own sovereign AI expertise. They could become part of the European Union efforts. They could do it themselves. It may not be leading edge, but will do more than a reasonable job. Many western countries have enough researchers and technical competency to match the skills from the USA or China. What they have is an economic framework to invest heavily.

2.8 AI Drives Growth in GDP

The benefits of AI will be to drive growth in global GDP according to most commentators. Energy disruption, as detailed in Optimistic Storm Pt 1 is a new era in energy, and creates opportunities as energy cost tends towards zero with superabundant energy, bringing 800 million people out of energy poverty and reducing a major input to society. In the AI transport section we saw that personal public transport had geopolitical ramifications, it would save households thousands annually, and with a reduction in accidents, save more than $30 billion a year or about 2% of GDP for Australia. OECD countries with higher road toll such USA would be higher.

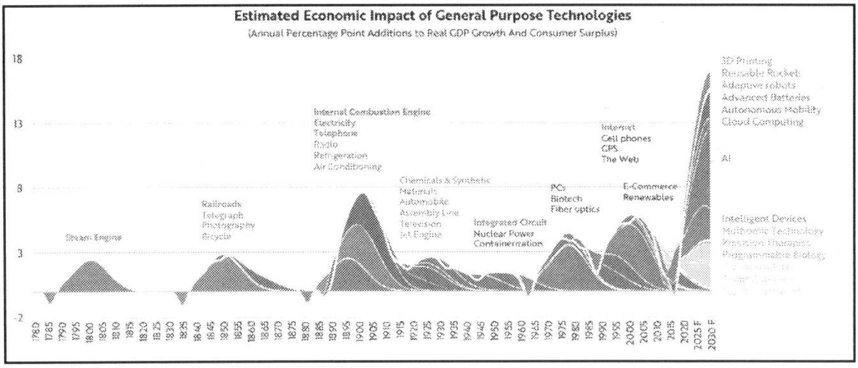

Figure 6 Economic Impact of General Purpose Technologies 1780 to 2030 Ark Invest 2024 Big Ideas

Likewise, AI will disrupt industries including transport, knowledge workers, health, agriculture, food and finance. Figure 6 shows past disruptions that contributed to additional GDP growth as innovations moved from lab to general use.

> *Technological convergence will create tectonic macroeconomic shifts more impactful than the first and second industrial revolutions. Globally, real economic growth could accelerate from 3% on average during the past 125 years*

to more than 7% during the next 7 years as robots reinvigorate manufacturing, robotaxis transform transportation, and artificial intelligence amplifies knowledge worker productivity Ark Invest 2024 [17]

These disruptions will also have a flow-on effect to the financial sector. We looked at the investment companies are making in AI, in data centres, in software systems, and in autonomous vehicles and those companies are expecting above-average returns on their investments. The companies at the forefront of the AI technology are currently worth about $13 trillion today and could be worth as much as $220 trillion by 2030. [18]

2.9 Conclusion of AI Overview

AI is here today. AI, like solar and batteries, has been under development for 70 or more years and will play an increasingly important role in all our lives. As we will explore in subsequent chapters in health, transportation, labour, financial systems, food and robotics. AI is already deployed in most areas of our lives today, not just the visible, such as search engines, chat functions on websites, chat boxes, spam phone calls, text messages, and cyber hacking. AI tools for good are already in use by computer programmers to write code, check code, and shorten development times. From personal productivity to drug development.

If there is one takeaway from this chapter is that AI is an enabling technology, and not the disruption itself. Some fail to understand the rapid progress AI has made. AI is the dumbest and most expensive it will ever be. We have seen rapid increase in the software part of the AI revolution which is the development of LLMs. The next chapter will highlight how much computer chips have advanced the computing needed to run these AI models – from training to actual use. AI performance doubles each six months. While the initial development has been USA-centric, the availability of open-source models and the plummeting costs of computing mean that China, Europe, Australia, or any country could free itself from the evil tech of USA billionaires and have its own AI systems, as China has demonstrated. They may be a year or three behind the leaders, but they enable geopolitical independence.

Learnings from An Overview:

- AI is here, accelerating, and can't be put back in the box.
- Digital intelligence is closing the gap with human intelligence – but consciousness and emotion remain open questions.
- The AGI race isn't USA-only: China and open-source models mean most countries can build sovereign AI capability.
- AI is forecast to drive a step-change in global GDP growth, on a scale beyond past industrial revolutions.
- AI performance is doubling roughly every six months – software gains are compounding alongside hardware gains.
- AI is an enabling technology, not the disruption itself – the next chapters show how it cascades into chips, energy, jobs, and more.

3 Chips for AI

The Role of Chips for AI:

- Computer chip innovation is driving the pace of AI improvements.
- Data centre investment is increasing.
- AI is a profitable business.
- What are the limitations on the expansion of AI?
- The economics of a single AI data centre – what it costs to build, and what it earns.
- From data centre to device – how inference chips bring AI to phones, vehicles, and robots.

3.1 AI is Getting Smarter

We can say with certainty that AI is dumber today than it will ever be. In the last chapter, we looked at the software part of AI. In this chapter, we will review the computers needed for a) training and b) inference. They are not separate, because both contribute to each other. The combination is much greater than the individual parts. Ark Invest predicted in 2024 [17] that:

- The major cost of training will decrease exponentially.
- The cost to train a model will fall.
- Training computers will increasingly use large data sets.
- Inference computers will use neural nets created.
- Most models will eventually run on modern mobile phones.

Declines in cost, increases in intelligence, and improvements in software have been staggering. We will look at past cost declines and then examine some AI performance improvements. Computing speed has followed Moore's Law for the past 50 years, which states that transistor counts on a chip will double every 2 years at about the same cost, with about a 50% increase in performance. It is not actually a law, but we can see the impact below of a supercomputer in 2006. In 2006, IBMs Roadrunner Supercomputer, with 17,000 chips (CPU) had a computing performance of 1 tflop (trillion floating point operations per second).

Figure 7 A 1 tflop computer in 2008 and 2020 (iPhone)

Figure 7 shows this à supercomputer in 2006 that took up a basketball court of space, and one in 2020 (iPhone 12) in the engineers hand. Both have one tflops of computing performance. Note that the iPhone 15 in 2023 had 2.4 tflops. The cost of computing performance and software development has declined by 65% year-on-year from 2015 to 2020. This is not the whole story. The type of computer processing is also very important, and many fail to comprehend it. In the '80s and '90s, computers had central processing units (CPUs), and with improved chip design, they enabled faster mathematical calculations. Graphics Processing units (GPUs) were developed, and users used them for applications such as bitcoin, which demands working with complex mathematics.

AI utilises these GPUs, but new chips have been designed for the mathematical calculations required for huge matrices, which can have hundreds to millions of rows and columns.

Because AI does most of its calculations on 3D vectors and complex arrays, those GPUs struggled. The latest chips are specifically designed to be much, much better to do AI maths. They need blinding fast speed and a huge amount of memory, at the same time doing complex 3D vector calculations and with access to incredibly large data sets. As we explore the new range of chip designs, we are just getting chips that meet those needs. The progress has been rapid, but there is every expectation that the next generation, past those announced or shipped today, will be further optimised for AI, whether for training or inference.

3.2 AI Data Centre Chips

The performance of AI supercomputers has doubled every 9 months for the past 6 years. Epoch AI has multiple data points showing the increase in computing.

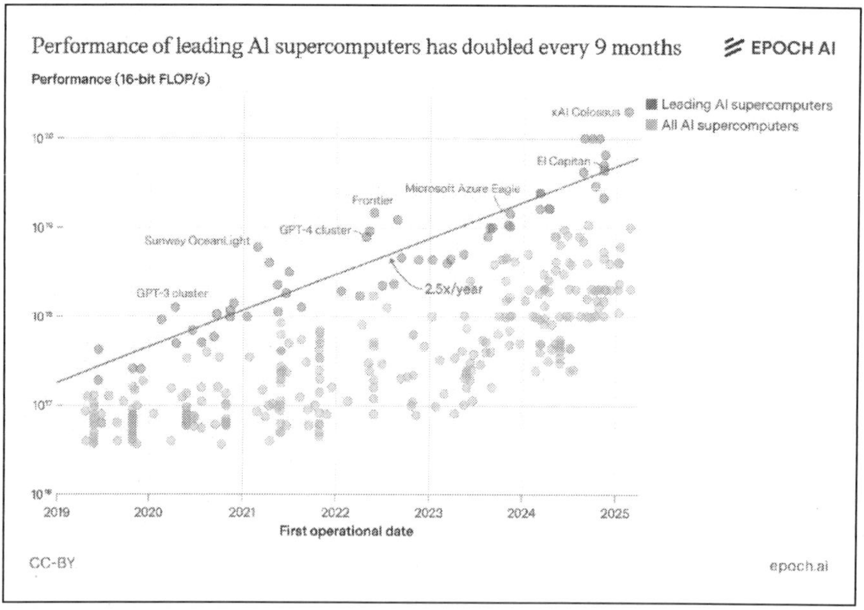

Figure 8 Supercomputer performance of AI computers doubling every 9 months

Figure 8 from Epoch shows the upward trend. The vertical axis measures computing performance of floating point operations per second. A teraflop is one thousand billion operations. The first teraflop computer, ASCI Red, was made in 1996. It was bigger than a basketball court. Size matters! A single modern Nvidia AI H200 Blackwell AI chip has a performance about 3,000 TFLOPS.

XAi built a supercomputer at its Memphis Colossus data centre has over 200,000 H200 chips for over 10^20 FLOP/s. This number is 100 billion billion, or 100 quintillion, which is a lot. Even that will be surpassed by another thousand times. When it was built, [19] it was eight times larger than HP's largest supercomputer in the world (1 quintillion FLOPS) or Aurora (2Q). Tesla claims it is no longer compute-constrained for its AI. It is building more data centres at its Texas site

Specialised Chips. Several companies have already developed chips and computers specifically for AI computation, including Nvidia, Tesla Dojo, Xilinkx and Cerebras. AI chips are still in an early phase, and we can expect to see new chips and new systems over the next decade.

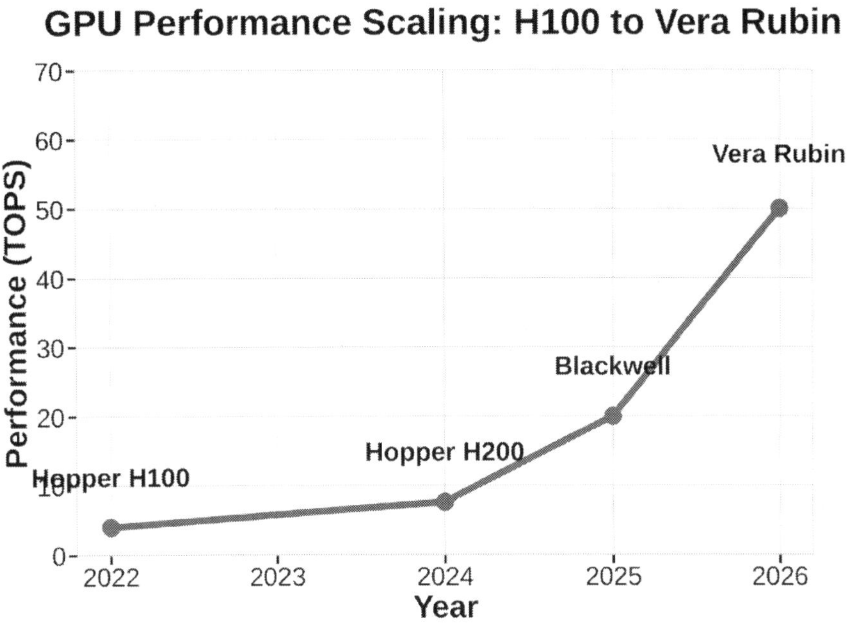

Figure 9 Improvement of Nvidia data centre computing

Figure 9 shows this relentless increase in AI speed. Nvidia, in particular, is pursuing a relentless pace of improvement, and CEO Jensen drove the point home by discussing how these innovations have made him the "Chief revenue destroyer". Jensen illustrates this by noting that Blackwell achieved a performance gain of up to 68 times that of Hopper, resulting in an 87% cost reduction.

Nvidia has a new AI computer called Rubin due in 2026/2027 to drive even more performance gains – 900 times that of Hopper, for a 99.97% reduction in cost. These chips come with all the other hardware to slot into a rack in server rooms. AMD and Intel have been lagging behind Nvidia in capability, and Intel has changed direction to develop new AI focused on cost-effective AI chips. [20] Intel claims its new Gaudi 3 is "not catching up" to Nvidia's latest GPU in head-to-head performance, but that the chip is well-suited to enable economical systems for running task-based and open-source models on behalf of enterprises, which is where the company has "traditional strengths." This is the "inference" chip segment, but in a data centre.

3.3 AI Inference Chips

Inference chips can run LLM models at the edge of the network on laptop computers, mobile phones, robotic vehicles, and human-robot systems.

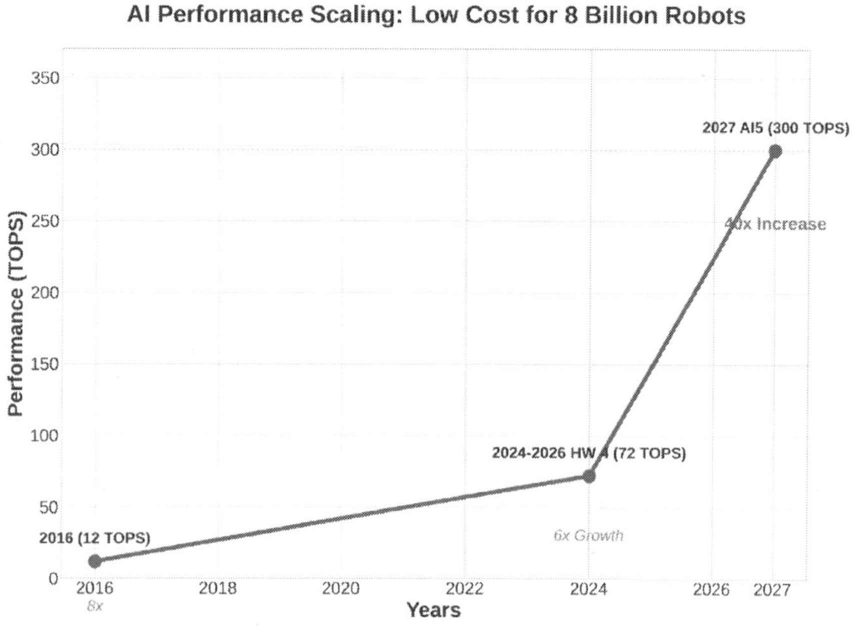

Figure 10 Inference Chips - running AI on mobile devices

Figure 10 shows the increase in performance of Tesla's hardware 3, hardware 4 and their announced AI5 chip, due in their vehicle and robot products from 2027. Inference is a rapidly expanding area of chip design. Tesla first developed their own HW3 in 2017 when it moved from Mobileye to their own hardware. Those chips were about 1 TOPS. HW4 improved performance to about 72 TOPS and their newly designed AI5 one about a 40x increase. Nvidia has also released its Orin and Jetson chips for car and robotic manufacturers. Every computer and phone manufacturer is looking for these devices, and low cost and high performance are important if you are after a 6 billion phone and a 8 billion robot market.

3.4 Software Plus Hardware

Along with the increase in computing hardware, we see improvement in software development. Costs are going down but performance is going up. For example:

- ChatGPT 4 reportedly cost over $100 million to train using previous chips. It is now performing at a level surpassing human expertise. The larger the model, the more it will exceed existing models. OpenAI, xAI, Meta, Amazon, Deepseek or other major model vendors are planning and continuing to release new models.
- Developing an application using those core models. Today, the training cost for a moderately sized, decent model is around $5 million. By 2030, training a new model might cost less than $500 and some have even said less than $50. Any business can invest in creating its own model for its operations, and do so in spoken English.
- The impact of these new chips are showing up in the performance of the models. It is complicated in that software is also undergoing optimisation. When XAI released Grok 4 in June 2025. Grok 4 scored 45% on the "impossible" AI test. Humanity's Last Exam features 2,500 expert questions across over 100 disciplines. GPT-4 gets 3%. Gemini gets 21%. Grok 4 hit 45% - more than double the previous best. Think PhD-level questions in physics, medicine and philosophy and xAI aims to release a major model every month and we see the same release schedule from all the major LLM foundation models.

Advances won't be all easy, though, as IT is littered with poorly executed software deployments. Queensland Health in Australia upgraded its systems for over $400 million, but the ensuing deployment debacle cost them more than $1.25 billion in payroll disputes. The Australian government implemented a RoboDebt program poorly, and it went horribly wrong, with costs far exceeding any benefits, let alone the implementation costs. Lessons don't appear to be learned, and daily we see companies announce big AI projects and fail. For all of those failed projects, dozens of others are successful.

3.5 AI Data Centre Gold Rush

Investment in new data centres and computers is enormous – both for AI itself and for AI applications, including robotics. Estimates are that over 1,500 new data centres are currently in various stages of planning and construction across the USA. Globally, industry groups are tracking over 500 large "hyperscale" facilities and more than 800 specific sites actively under construction. [21]

McKinsey research indicates that by 2030, data centres are projected to require $6.7 trillion worldwide to keep pace with the demand for computing power. Data centres for AI are projected to need $5.2 trillion in capital expenditures, while those powering traditional IT applications are $1.5 trillion. [22] In 2025, spending projections for four AI companies include:

- Alphabet $75 billion in capex, up from $33 billion in 2024.
- Microsoft $80 billion for training AI models and deploying AI and cloud-based applications around the world, up from $46 billion in 2024.
- Amazon $100 billion to reinforce AWS up from $19 billion in 2024.
- Nvidia and xAI have joined a $30 billion fund dedicated to data centres operated by BlackRock and Microsoft. [23]
- Tesla has invested over $10 billion for autonomous cars and robots.
- The capital investment in 2026 alone is calculated to be over $750 billion.

This level of investment will boost AI capabilities and allow customers to deploy their applications, as we will look at sector by sector in the next section. It's an "arms race," with all these big corporations betting heavily on the market. Ark Invest, in its Big Ideas 2025 [17], says the costs of AI fall primarily due to architectural improvements in AI systems, and that performance per dollar of AI compute is expected to improve more than 1,000-fold by 2030. By then, Ark expects that computing performance will have doubled 64 times since the advent of the integrated circuit. The combination of increasing computing power and improved software supports the early prediction that AI will improve at an annual rate three times faster, far exceeding Moore's Law for computing.

3.6 AI – Why the Gold Rush?

Some consider the current boom in AI a passing fad, citing past mining and internet booms, but fail to account for the impact AI will have on the labour market. If you assess the various components of global GDP.

- Total world GDP = $126 trillion
- Labour = 50% of total GDP
- Current labour cost of $13 per hour. (Discussed further in the robotics chapter).
- Labour cost of AI-powered humanoid robot = $0.50
- Market opportunity is over $50 trillion.
- With 8 billion robots, that is over $6,000 per year for over 5 years.

Companies and governments have these figures in mind and do not want to miss out on some of that market share.

3.7 AI Data Centre Economics

Getting data to understand AI data centre economics is generally commercial in confidence and hidden, but with Anthropic and SpaceX/XAI floating on the sharemarket in IPOs, then some data has been exposed. The cost to build one of these monster data centres is listed below. A data centre is just a large building that houses computers and requires electricity and water for cooling. We will do the numbers for a 1 GW data centre, and this information comes from Chase Lochmiller, CEO at Crusoe.

- A 1GW electricity plant powers about a million homes. Eg San Francisco, or an aluminium smelter.
- Each 1 GW data centre will have hundreds of thousands of chips.
- Total Cost of Data centre = $60 billion

- The building and electricity = $19 billion.
- Over $40 billion of IT equipment inside.
- Half (47%) is the computer chips – the balance networks, data storage and backup.

Revenue from a 1 GW AI Data Centre.

- It is a $15 billion annual revenue opportunity. How do we know that?
- Anthropic leases one of Tesla AIX for over $1 billion a month.
- Open AI has said it costs $20 billion to operate a 2GW data centre and half is training, and the other half is revenue.
- Hence, the revenue opportunity is more than $20 billion.
- Revenues are not decreasing. OpenAI states its new system doubled revenue in a year.
- Pay back under 2 years with $30 billion in revenue.
- Chips last for more than 4 years, but could be used for inference rather than being the high-end chips needed for the latest training. The building and electricity are probably good for a decade.

As models become 100 to 1000 times more capable and training software becomes 2 to 3X better at training, training costs will fall.

- The revenue per watt has seen a 10 to 30x increase over the past 3 years and is likely to increase further.
- Everyone gets a bit of those increases.
- Nvidia passes much of the value of new chips on to others.
- The trend is for more intelligence per watt over time.

Tesla has floated the idea of putting data centres into space. At first glance, it appears hairbrained. But, a space-based AI data centre is comparable in cost to on-earth data centres, but may win on speed to market.

- Put the satellite into HEO so that it can generate electricity 24/7 with solar panels and the cooling requirement can be met with radiation. Transmission is with light, and while there is some latency, latency is not too important in training models.
- It costs about $100m to build Starship.
- If the cost of getting a kg up is $1,000, then it would cost over $25 billion to launch. The cost to launch may to $300 per kg, then it could come

down $7.5 billion, which is similar to the cost of a land-based data centre.

The argument is about scarcity on Earth versus speed and scalability in space.

- The costs for earth-based data centres are likely to go up. There is a scarcity of sites and electricity supply.
- The velocity to get to market is more important than the price when the return from a data centre is 2 to 4 years at most.
- Tesla will have a conflict between sending up more Starlinks and building space data centres; it is likely to be 2032 before we see any space data centres.

3.8 Conclusion of Chips for AI

We can see how the advancement of microprocessors (chips) has accelerated AI. Yes, AI software has leapt forward in leaps and bounds over 70 years, but it is the advancement of specialised AI chips that is revolutionising AI and helping it double in speed and performance every six months. With new chips and data centre systems, AI have orders of magnitude more capability now than it did a few years ago.

Conclusion of the Role of Chips for AI:

- Computer chip innovation is driving the pace of AI improvements with hundreds of times faster processing.
- Modern chips are 50 to 100 times faster than just 5 years ago.
- There is not end in sight for improvements.
- Tech companies are investing over $750 billion into data centres over the next five years.
- A single 1GW data centre can cost $60 billion to build but generate $15 billion or more a year – the economics explain the gold rush

---

4 Electricity & AI

Introduction to Electricity and AI:

- The world is electrifying rapidly.
- Data centre investment is increasing.
- What is the global supply of energy?
- Will AI help or hinder the electrification of everything?
- How much electricity do AI data centres actually use, compared to the rest of the grid?
- From hominids to data centres – a quick history of how human energy use has scaled.

One of the hot topics in AI is the additional demand for electricity to power the data centres and computers, and where will it come from? There have been wildly exaggerated claims and counterclaims in the media about how much additional electricity will be needed. It is said that more electricity is needed for AI than for "normal computing" because AI has to do very large matrix calculations, and that is extremely energy intensive, as we have seen in the previous section.

4.1 AI Demands Electricity

These modern AI chips, specifically designed for matrix calculations, are extremely power-hungry, even though their energy-efficiency improvements are remarkable, as we will see. However, when you have hundreds of thousands of them, they use megawatts of electricity. How many data centres are planned? As pointed out previously, about 1,500 new data centres are currently in various stages of planning and construction across the USA. Globally, industry groups are tracking over 500 large "hyperscale" facilities and more than 800 specific sites actively under construction. The energy demand is about 23 GW currently. The biggest demand for electricity is training those AI data models. New, customised chips from Nvidia, Tesla, and others, designed from first principles, are much faster and consume less power, as evidenced by their announcements. When Microsoft, OpenAI, Google, Meta, Tesla, or other data centres have over 200,000 AI computers, even at a few watts, the demand is equivalent to terawatts of electricity. Some information about electricity demand includes:

- Power requirements = aluminium smelters.
- Globally, today, data centres = 1.5% of total electricity demand.
- Data centres are concentrated in a few areas. They use a high proportion of electricity only in some places. E.g. 20% of electricity in Ireland and in Virginia, USA. In other areas are less than 1%, and even less outside the USA.
- AI data centres are expected to be 50% of new electricity demand in US.
- Key factor is how quickly a data centre can be built, and that is now a function of electricity supply.

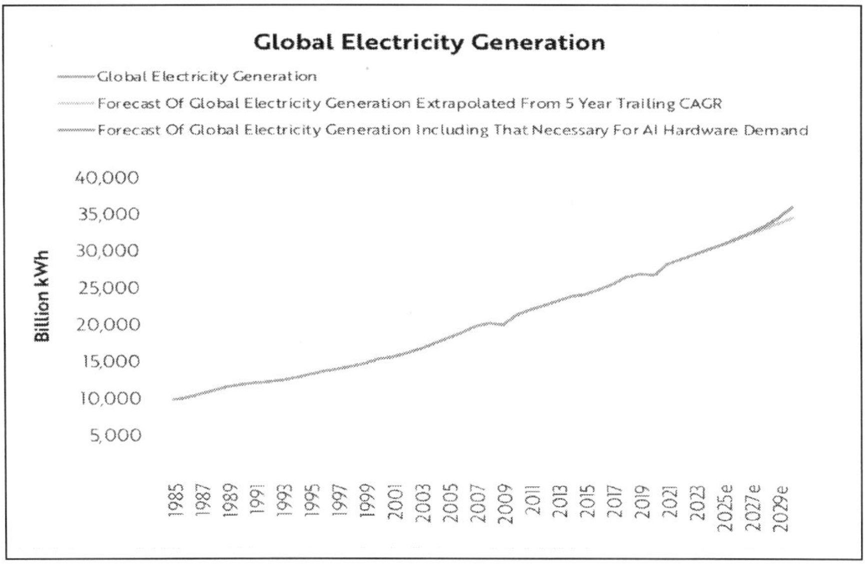

Figure 11 Global Electricity Generation from 1985 to 2029.

Figure 11 from Ark Invest shows global electricity demand through today and the forecast. A billion kWh is a TWh. Michael Liebreich in Bloomberg [24] points out that there are conservative projections, such as IEA, and much more aggressive projections. [22] The rise in electricity demand will be noticeable, but attainable says Ark Invest [12] who point out that the total new electricity supply needed is slightly more than what China added in renewable energy in 2024.

We need to understand that while the primary demand is for training, there is also substantial electricity demand for the AI inference component. Inference AI can be located in a data centre, on a customer's site, or on a mobile device. Companies are taking different approaches. If you use CoPilot in Microsoft Office, you are probably performing inference in the Microsoft data centre. If you have a high-spec AI computer, some of that inference will be local.

For a detailed understanding of electricity and energy, we should refer back to Optimistic Storm Part 1. To put electricity demands into context, we will revisit some basic energy requirements for our human civilisation.

4.2 Global Energy – A History

Let us begin with early hominoids, approximately a million years ago. Humans have extremely low energy requirements and are one of the most efficient mammals. The typical human diet of 2000 kcal/day is equivalent to approximately 100 watts, and our brain, although it accounts for only 2% of body weight, consumes approximately 20% of the body's daily energy. Energy was critical – if we did not have enough energy, we died. We, therefore, are naturally "lazy" and conserve energy as it is critical for survival. Fire discovery doubled energy availability, enabling us to consume more protein and more energy-dense foods. With more food, we did more work. We started using animals for work only about 6,800 years BC, and that meant many more kilowatt-hours of work. An abundance of energy saw the rise of civilisations. The next change was coal and oil, which increased our energy supply by ten to 100 times.

- Ada in sub-Saharan Africa uses about 0.18 MWh per year.
- Tracey in the West uses over 13 MWh per year in the USA and 6 MWh in Europe.

- Energy disparity between the rich and the poor is 100 times.

Many fail to understand the significance of the disruption caused by the transition to renewable energy and still cling to the fossil-fuel era.

- A single modern solar 450W panel generates about 1.8kWh daily or 600kWh annually.
- One PV panel produces three times more electricity than Ada uses.
- We only need 30 PV panels per person for the West (USA) to replace all the energy consumed.
- Note that we do need batteries to store electricity from when it is captured, to when we need it. Fortunately, the amount of batteries needed is now plentiful and cheap.

On a global scale, as the world transitions from fossil fuels to renewable energy, we will need less energy.

- Energy demand is about 166,000 TWh per year.
- We waste 60% of this primary energy in extraction, refining, transport and poor efficiency.
- Electricity supply is 30,000 TWh generated by fossil fuels and clean sources in 2026. It has risen as solar and wind overtook coal in 2025.
- 16,000 TWh was generated by coal and gas, with the balance clean energy (hydro, nuclear, wind, solar).
- Switching from fossil fuel to clean renewable electricity is the most effective and fastest path to decarbonisation.
- 66,000 TWh of electricity would decarbonise completely.
    - 15,000 TWh replaces oil burned for transportation.
    - 15,000 TWh is needed to replace fossil fuels used in industry, heating, and processing.
    - 10,000 TWh for removing energy poverty.
    - 10,000 TWh for growth in electricity demand.
- IEA suggests a 30% rise in electricity use by 2030, whereas others say there will be a 250% increase.
- RethinkX, in their Stellar Energy for superabundant electricity, suggests that electricity can increase by over 300% and bring nations and billions of people out of energy poverty. [25]

Numbers are higher in the USA due to high industrialisation.

- USA electricity consumption = 4,178 Terawatt hour (TWh).
- USA uses about 15% of global energy.
- Data centres use about 2.5% of the USA electricity, having grown from the 2,000s of about 1%.
- AI demand increase could be as much as 100GW extra by 2030, with each data centre needing about ~26 TWh of electricity.
- Nvidia points out that electricity efficiency for their chips has come down 50 times in the past five years. Increased efficiency is not cancelling out all the extra electricity demand, but it meeting some of the extra demand.
- Some suggest many electricity grids are unsuited to the green revolution. E.g. Alimeter More grids enable wider resilience to a more intermittent electricity supply. [26]
- Others disagree and say data centres should have substantial on-premise generation and storage.
- Renewable energy, especially solar and batteries, can ramp up fast enough to meet the increase in demand, whatever the actual amount.

4.3 Electricity for Computers

The electricity demand is large not just for the chips, but also for the data and energy for cooling. The average Power Usage Effectiveness (PUE) – the ratio of total power used in a data centre to the power used by its servers – dropped to 1.5 in 2021 from 2.7 in 2007, with the best data centres now delivering PUEs as low as 1.1. The excess heat is at a low temperature and mostly dumped. To calculate how much electricity will be needed, we will explore how computers use electricity.

- A Nvidia H100 chip's power demand is ~700W or about the same as a small refrigerator.
- The new B200 Blackwell, with 8GPUs, 2 Intel GPUs, is 14kW max with compute of 72 petaFLOPS training and 144 petaFLOPS inference. That amount of electricity is about 20 times what an Australian household uses.

- Data centres do use a LOT of electricity. A Microsoft data centre employee says they cannot have more than 100,000 H100 servers in one state in the USA without exceeding that state's power supply capacity.
- A 20,000 GPU-equipped data centre, along with petabytes of data storage, would need a power of 300MW and achieve 7 exaflops. That is what a small city of 200,000 would use.
- For 100 AI data centres, demand would be 2,600 TWh. That is approximately 10 times Australia's annual power consumption (210 TWh).
- Meta, Microsoft, and xAI are all looking to co-locate with under-utilised power plants, and some have even proposed using their own dedicated nuclear power plant.
- In June 2025, Amazon announced a $20 billion investment in Australia by 2029. [27] With a target for 82% of the Australian electricity grid to be powered by renewable energy by 2030, and three new solar farms, Amazon can have plenty of clean electricity to power its centres. This press release also stated that, according to the Australian Government's Department of Industry, Science and Resources, AI and automation are expected to contribute up to $400 billion annually to Australia's GDP by 2030. Believe it or not.

While people are talking up the increase in demand, computers and chips have become much more efficient. Nvidia says it has reduced power consumption by 40x. Koomey's Law states that the energy efficiency of computers doubles every 18 months for six decades, almost exactly matching the improvements in computer performance predicted by Moore's Law.

Tesla adopted a strategy of using very low-power inference computers in their cars. Hardware version 3, 4 or 5 computers in the car draw less than 100W. 100W is about the energy used per 1 km of driving. To compensate for the vehicle's limited inference computing power, Tesla requires significantly more computing power on its training supercomputers.

4.4 Conclusion of Electricity and AI

In conclusion, the electricity demand by AI will increase. There is much media speculation about how that extra electricity will be provided, and it is very clear it will be from wind and solar. The ongoing reduction in costs for renewable energy and batteries, short construction times for renewable energy projects, and the efficiency of AI computation mean electricity is unlikely to be the constraint on the continued deployment of AI. Nor will it break the electricity sector. To conclude, as we discussed in Part 1 of Optimistic Storm, Ada and the 800 million others like her will need considerably more electricity to transition from energy poverty to energy abundance than the electricity needed to power AI.

***

Conclusion to Electricity and AI:

- Demand for electricity by AI is high.
- The electricity supply will mostly come from wind and solar, backed up with batteries.
- At most, AI will use about 2.5% of energy globally.
- Improving chip efficiency is absorbing much of the extra demand – Nvidia says power use per chip has fallen 40-fold.
- Electricity won't be the constraint on AI deployment – nor will AI break the electricity sector.
- The biggest demand is to bring 800 million people out of energy poverty, and that won't be concentrated in the West but distributed across Africa and Asia.

***

5 Risks of AI

Introduction to the Risks of AI;

- What risks does AI have? Weaponisation, runaway superintelligence, growing inequality, and external threats.
- The doomist versus accelerationist divide: who is worried, who isn't, and why.
- How AI could be weaponised – cyberattacks, bioweapons, election manipulation, and autonomous weapons.
- How do we mitigate these risks, and who decides?
- The growing inequality risk – will AI's benefits flow to the few, or to society at large?
- Low-probability, high-impact external risks – from electromagnetic events to civilisational resilience.

Most experts in the AI field, including the father of AI, Nobel prize winner Geoffrey Hinton [55], point to the discourse around superintelligence or AGI. There is a fundamental divide between two schools of thought: the doomist camp anticipates catastrophic outcomes from the emergence of superintelligent AI. The accelerationist camp of Eliezer Yudkowsky [28], founder of the Machine Intelligence Research Institute (MIRI), and Yann LeCun [57], Turing Award winner and Chief AI Scientist at Meta [58], are two who consider such concerns overblown or misguided. Others have warned about the risks. Elon Musk [29] warns about the dangers, stating that there is a 10% to 20% risk of AI going wrong, yet continues to raise billions for xAI. There are many types of risks, but essentially they fall into the following categories.

- People will weaponise AI.
- AGI risk. AI becomes superintelligent and surpasses humans.
- Risk from growing inequality. Value will accrue to the tech billionaires. The rich get richer and the poor get poorer.
- External violent Electromagnetic Radiation

5.1 Risk from Weaponisation

AI improvements won't stop. It is too good for health, education, food, business improvement and customer service but the risk with robotics, bioweapons and cybersecurity will get worse.

- Phishing attacks in cybersecurity were 1200% higher in 2024 than in 2023, as ChatGPT allowed cloning of voice and data and closer targeting of victims. AI can scour millions of lines of code, searching for vulnerabilities that can be exploited or leveraged. In June 2025 a record-breaking data breach exposed 16 billion passwords to the world, including user credentials for the likes of Facebook, Google, Apple, and tons of other places. Some commentators were quick to call it the largest password leak in history, and in terms of raw records exposed, that's mostly technically true. However, these records did not come from a single breach – or even a new breach. Instead, they came from many smaller ones. AI is likely to make these compilations and hacks easier. AI will lead to cyberattacks. It will get worse as there will be more data breaches. We can assume some techniques are yet to be invented. The only benefit is that AI is already protecting systems from cyberattacks and we will see a race of good AI versus bad AI.
- Molecular Biology Warfare. AI can now create new viruses as we will review in Health. For a few million dollars a rogue company or state could fund a program to develop a novel disease. When the bioweapon is released, the world will struggle to find a vaccine or cure before millions or billions are dead. The Bulletin of Atomic Scientists point out risks in detail. [30] Science had previously reported an experiment at MIT, where researchers asked a set of undergraduate students to use an LLM to create and order a dangerous virus capable of unleashing a pandemic. Within an hour, the chatbots had suggested a list of four potential pandemic pathogens. In some cases, the chatbots pointed to

genetic mutations reported in the literature to increase transmission. The LLMs described how viruses could be created from synthetic DNA using reverse genetics and provided the names of DNA synthesis companies that were judged unlikely to screen orders. They also supplied detailed protocols and guidance on troubleshooting, and recommended that anyone lacking the necessary skills could engage with other organisations to overcome these challenges.

- Corrupting Elections. AI enables the segmentation of voters. AI can mine all available data about individuals, link that data and then target electors individually. Current social media systems are already adept at performing data analysis and targeting users with relevant advertisements; however, AI offers a whole new level of functionality and cost-effectiveness.
- Creating echo chambers – making people angry but feeling righteous. You can see this with mainstream media, and there are many examples. E.g. violence in Victoria, Australia. Most malicious political actors know how this works, and the USA is a prime example.
- Lethal autonomous weapons. Big, powerful countries can invade smaller countries. The only problem with this is that war has an unpleasant side effect, in that dead soldiers coming back in body bags cause very severe anti-government backlash. We saw this with Vietnam and Iraq. The people can change the political climate changes from support for war to opposition against it. AI reduces both the direct cost of war and the political opposition. The race to perfect remote killing is being run at a furious pace. See the Drones section for more details.
- AI will replace people in all sectors and in all industries. Companies and governments may use this to weaponise against other companies and governments. The duopoly of Coles and Woolworths in the food sector show they target competitors to drive them out of business.

5.2 Risk from SuperIntelligence

Hinton [31] states that there is no way to eliminate the superintelligence/AGI. MIT details some of the risks [32] and they boil down to the challenge of controlling an agent that is substantially more intelligent than us. The battle will be for control over humans by ultra-smart machines.

- AGI removing itself from the control of human owners/managers.
- AGIs being given or developing unsafe goals.
- AGIs with poor ethics, morals and values.
- Inadequate management of AGI.
- Existential risks of AGI.

Hinton and others argue that a super-intelligent robot could create a virus that could wipe out the entire population or a subset within 10 to 20 years, with no way to prevent it. We previously discussed the risk that a human or state actor could do so, but in this case, would an AGI robot decide to take this course of action? It knows humans are destroying the Earth's ecology with emissions and agriculture, and that humans wage war against each other. What if it were decided that the best solution would be to reduce the population back to 1 billion and bring the world to its sustainable carrying capacity? Check out the population chapter. What will superintelligence do to our children and grandchildren?

5.3 Risk of Growing Inequality

In the population chapter we talk more about inequality.

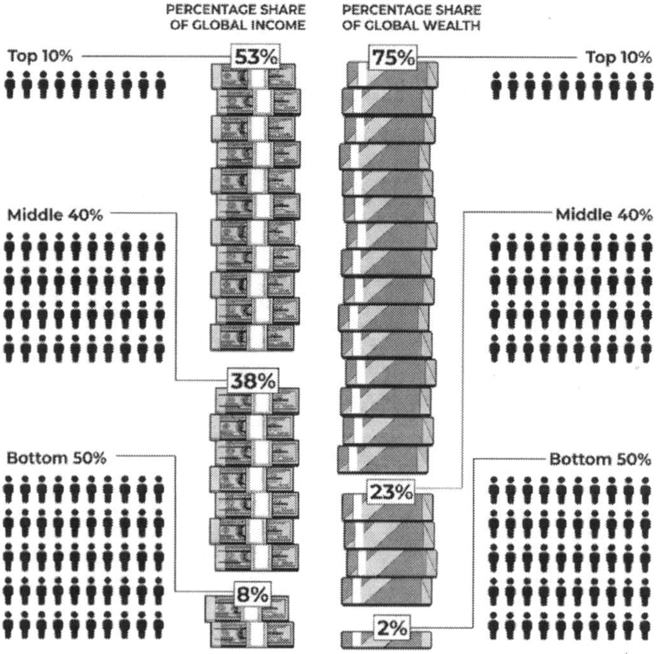

Figure 12 Income and wealth

Figure 12 shows that for income, the top 10% have 53% of the global income. But those 10% have over 75% of the wealth. The bottom 50% earn just 8% of the income and have just 2% of the wealth. A risk of AI is that most benefits from AI will flow to those 12 billionaires who own 50% of the world's wealth. What happens to those who lose their job and their lives are disrupted? Do we actually trust a fair sharing of benefits?

5.4 Electromagnetic Radiation

It is low risk, and many dismiss the risk, but violent electromagnetic radiation from outside our solar system is a possibility. External events that we have dismissed such as violent volcanic eruptions or major climate change events could bring disaster to electrical and data networks or AI data centres. Elon Musk says that they need to set up a planetary outpost to be able to survive this low-probability, high-impact event. Do we have the disaster recovery systems in place? Humans have been around for a million years, but we have no planet B in place.

5.5 Conclusion of AI Risk

We discuss the benefits of using AI in various chapters – autonomous driving, food supply, development of vaccines and medicines and diagnostics changing our lives for the better. The loss of jobs is going to create enormous social pressure as we discuss in both the services and robotics chapters. Now, how do we stop AI from making even more destructive weapons?

The risks from AGI are real and present, and present a real threat to human survival. What do we do?

- Halt future development?
- Pause the progress until we have appropriate policies in place?
- Put checks and balances in place now, before we get AGI.

We know that outrageous greed can destroy humanity unless we have a spiritual/moral approach to prevent that greed. Greed in Russia is slowly destroying the culture and fabric of society. Greed and power keeps many

African states in conflict and poverty. We cannot eliminate greed – as many civilisations before us have attempted and failed.

Can we make a difference to the advancement of AI and robotics? We can, and we must target our politicians and lobbyists to ensure that this change is positive across most sectors. We should be enthusiastic about the areas that will benefit from AI. We need to apply our efforts to eliminate the drudgery and time wasted in healthcare so that we can focus on more effective care. We want teachers to spend more time on education and less on crowd control. We need workers in service industries to be more efficient and targeted, allowing us to pursue our passions with a system that benefits everyone in society from the changes to come.

What we don't want is for the benefits to accrue to the wealthiest tech billionaires. As we see in the weaponisation of machines, we don't want wars.

What did we learn from risks from AI:

- Misuse of AI is likely – weaponisation, cyberattacks, bioweapons, and election manipulation are already underway.
- Superintelligence risk is real and cannot be eliminated, only managed with guardrails put in place now.
- Benefits are likely to accrue to the wealthiest billionaires unless we change the narrative.
- We need checks and balances in place before AGI arrives – not after.
- We have the tools today to make AI a net positive – eliminating drudgery in health, education and services.
- The choice is ours: a future of shared abundance, or one where AI deepens existing power imbalances.

---

6 AI Disrupts Service Jobs

How Will AI Disrupt Service Jobs:

- White-collar or service jobs are 80% of jobs in the West.
- The Human Agency Scale: a framework for which jobs AI automates versus augments, and how soon?
- AI will initially start doing tasks. Not jobs.
- How will tasks and agentic AI impact jobs? In what time frame?
- Government Jobs – at what risk?
- What are the implications on a global scale?

AI offers both excitement and concern regarding its impact on job creation and disruption. The effect of AI on employment will vary across different countries and industry sectors. What are the trends and potential outcomes? AI is a truly disruptive technology that will accelerate disruptions in energy, transportation, health, labour, food, and agriculture. In the introduction to ths book we summarised disruptive technologies and showed how different futurists talk about disruption, causes and consequences. RethinkX (Dr Tony Seba) defined eight technologies for three disruptions [33], whereas Ark Invest defined 14 technologies but identifies five disruptions. [34] Both show AI is a key technology enabling disruption. Generally, a disruption needs multiple technologies.

In this chapter we will examine how AI will impact jobs in the service sector and in the next will look at specific sectors and predict what will happen to them. The service sector is sometimes called white-collar or knowledge workers. AI will be the primary technology in the disruption. For the manufacturing and blue collar sector, discussed in the robotics sector, the consequences of human labour being displaced by machines have enormous consequences on geopolitics, both trade and war. Many advances in health are enabled by AI, as we will explore. Agriculture and food will be disrupted partly by AI technologies, but additionally by the energy revolution in play. While these transformations are underway, they are "below the radar" for most. This book attempts to help you understand how profound the changes may be, and what timing. Timing is less certain because we are at the beginning, maybe 1% or even less.

6.1 Scale of AI Job Disruption

The impact of AI on jobs is not a one-size-fits-all scenario. While some jobs may be replaced by AI, new roles will emerge, requiring human skills that complement AI capabilities. Adaptation and upskilling will be crucial for individuals to remain relevant in the job market.

AI Preparedness Index and employment share in high-exposure occupations

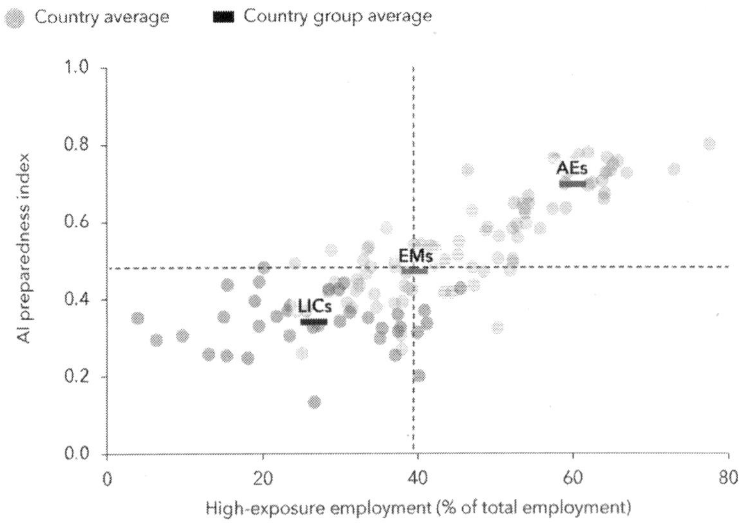

Source: Fraser Institute, ILO, International Telecommunication Union, United Nations, Universal Postal Union, World Bank, World Economic Forum, and IMF staff calculations.
Note: Plot reflects 32 advanced economies, 56 emerging market economies, and 37 low-income countries. Dotted reference lines are derived from AI Preparedness Index median values and high-exposure employment.

IMF

Figure 13 AI preparedness for Low-Income Countries, Emerging Market Economies, and Advanced Economies.

Figure 13 from an IMF assessment of over 32 countries determined that over 40% of total employment is at over 50% risk. They also concluded that most countries are unprepared for the changes.

The economic debate continues to be dominated by two camps: one claims that AI will replace jobs, but that overall, more new jobs will be created. The other camp sees the opposite: AI will replace significantly more jobs than it creates. The impact of AI on job creation and disruption will vary across countries and industry sectors. While there may be short-term challenges, the long-term benefits of AI adoption can outweigh the initial disruptions. Governments and organisations should focus on providing education and training programmes that equip individuals with the skills needed to thrive in an AI-driven economy. Collaboration among industry, academia, and policymakers is essential to ensure a smooth transition and minimise the negative consequences of job disruption. A proactive approach that focuses on education, upskilling, and collaboration will be vital in maximising the positive impact of AI on the global workforce.

6.2 Human Agency Scale

To determine whether AI will disrupt jobs, which jobs and when, it is helpful to have a framework that allows for a detailed examination of which jobs are at risk, why those particular jobs are vulnerable, and how soon AI will disrupt them. Shao et al [35] provide a helpful definition and use a "human agency scale." They looked at the future of work with AI agents in AI hyperautomation. They interviewed 1,500 workers and 52 AI experts across over 844 tasks spanning 104 occupations. Most raised concerns about job displacement, diminished human agency, and overreliance on automation. In the following Figure 14, we look at the authors' framework of a "Human Agency Scale (HAS).

HAS ranges from H1 to H5. With H1 or H2, AI replaces human capabilities, and the AI takes primary responsibility for task execution with little or no human oversight. At Level H4 or H5, the human drives the task completion, and the AI agent requires human input to finish the task successfully. The human directs the task completion. In the first case, AI's role is automation; in the second, it is augmentation.

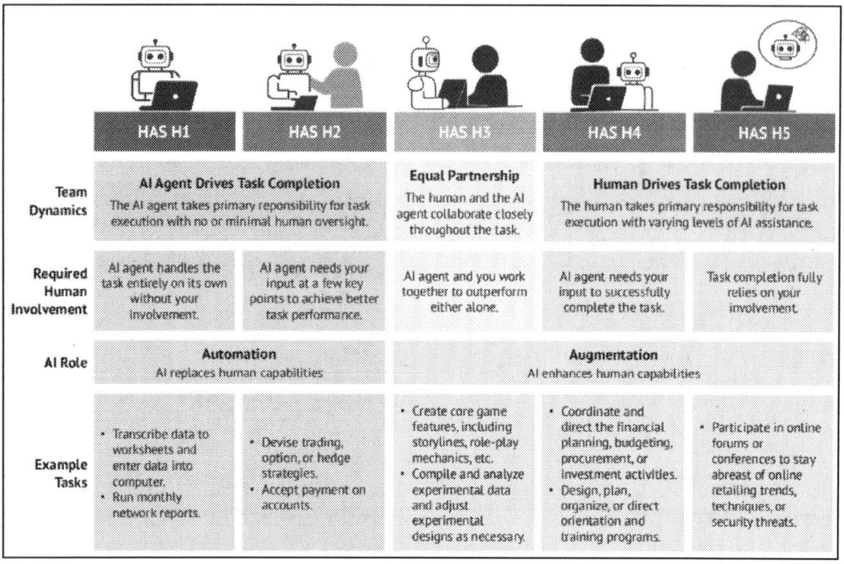

Figure 14 Human Agency Scale from Future of Work with AI Agents: Auditing Automation and Augmentation Potential

Most industrial robots are fully automated (H1). That automation does everything. In contrast, humanoid robots will be capable of both full automation and some augmentation. Driving with advanced driver-assist systems (ADAS) today is about H4, whereas an autonomous vehicle is entirely automated at H1 or H2. Humans dislike the intermediate steps of H3 and H4 because they don't know if they can trust them or not. We see this dislike if driving a modern car, and the car insists on following the lane-following control at H3. Providing sales assistance in a dress shop is likely to be augmentation (H4), as the salesperson uses AI to display colours or assist with fitting, and to process payments, but the stock ordering, tracking inventory, and restocking are likely to be HAS2 and HAS3.

6.3 Temporary Augmentation

When we consider the human agency scale, we see that on one side is augmentation of jobs, and at the other end is replacement. The questions we don't know are how much augmentation there is and at what stage job replacement will occur. We know today that AI can replace a task. We can see this process underway today.

This process of Agentic AI, where multiple tasks are coordinated, is not just augmentation but replacement of real jobs. If you are a small business, AI could enhance productivity. A large business with multiple tasks for multiple people means replacement.

6.4 Service Jobs 4x Blue-Collar Jobs

Globally, the trend is that white-collar or service jobs are increasing for several reasons. The shifting tide is due to:

- Economic. Most Western economies have transitioned from a manufacturing-based economy to a service-based one, with a large proportion of jobs in finance, health, education, retail, and other service sectors.
- USA, Australia, Germany and European countries are all now over 70% to 80% service jobs.
- Automation and technology advancements are reducing blue-collar jobs, particularly in manufacturing, leading to job displacement in some areas.
- Globalisation, combined with offshoring, has led to the relocation of manufacturing jobs in high-wage economies to countries with lower labour costs.
- Aging of the blue-collar workforce.
- A decline in the number of young people entering trades.
- Privatisation of large government industries. E.g. Telecom Australia took on thousands of apprentices. With privatisation, the focus moved towards shareholder value and cost-cutting, leading to a reduction in the number of apprenticeships offered.
- Demand and Supply of Blue-Collar Jobs – There is a shortage of skilled workers in blue-collar fields, such as construction and trades. E.g. In Queensland, Australia, with the Olympics scheduled for 2032, it is estimated that there is a shortage of 30,000 to 50,000 construction workers competing for projects in rail construction, road construction,

Olympic infrastructure, mining, house building, and renewable energy.

6.5 Number of Jobs At Risk

Different organisations have different estimates of how many jobs will be augmented or lost. The IMF [36] states AI will impact 40% of jobs overall. Over 60% of jobs in advanced economies will benefit from AI integration, while less than 20% of jobs in low-income countries will.

> *Historically, automation and information technology have tended to affect routine tasks, but one of the things that sets AI apart is its ability to impact high-skilled jobs. As a result, advanced economies face greater risks from AI—but also more opportunities to leverage its benefits—compared with emerging markets and developing economies.* IMF

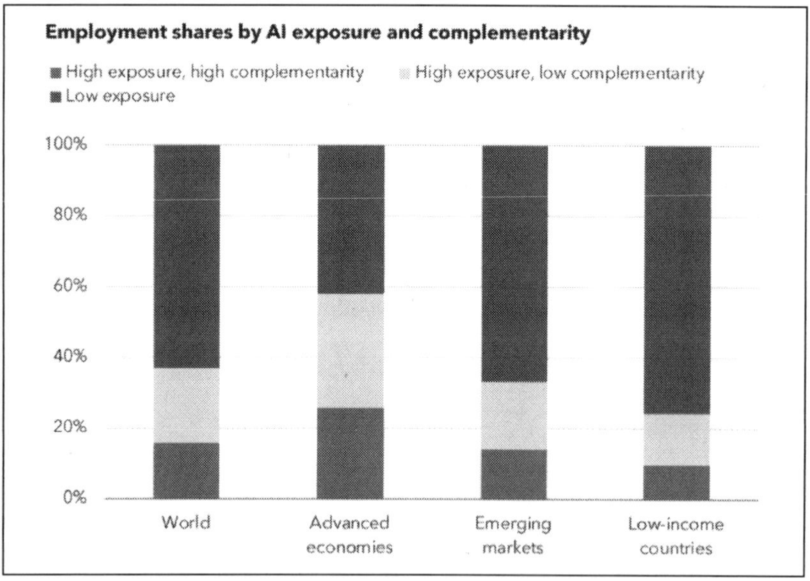

Figure 15 AI's impact on jobs various economies.

In Figure 15, the Word Bank outlook is more aggressive than the IMF. They project 65% of all jobs will be displaced by robots and AI by 2035 in advanced economies including the USA. Keep in mind that there are multiple separate sectors for AI. Job disruption differs across these sectors, with varying degrees of augmentation or substitution.

- Services – white collar or knowledge workers
- Labour – blue collar by robotics
- Transportation – robotics plus specialised equipment.

Some commentators say 10% of jobs that robots do are already being done [17]. That number is almost certainly an underestimation.

6.6 Government vs Private Jobs

Governments employ a large number of people. Will AI increase their productivity or will governments look to reduce staffing levels and introduce more AI for services? How many jobs are we talking about? The number of public servants in Australia [37] is 2.5 million. Their wage bill of $232 billion is 22% of all wages. Wages for the private sector are 78% of all wages paid.

- Commonwealth = 0.36 million. 22% are military personnel.
- State government = 2 million (education and health)
- Local government = 0.2 million (3 R's = roads, rates and rubbish)

The majority are service jobs. The three largest sectors make up 90% of wages and include education (0.7 million employees), health (0.6 million), and public administration and safety (0.8 million). Comparing different countries [1] shows similar numbers. The public sector in Scandinavian countries accounts for around 30%, whereas most Western countries are closer to 20%. The USA is lower but still has 23 million public employees across federal (2.9 million), state (6 million), and local government (15 million). Public employment accounts for approximately 14%, while private employment accounts for 70%. You have to be careful about making comparisons, as much of this difference between countries stems from the fact that public services, such as healthcare and education, are provided by the government in some countries and by private entities in others.

6.7 Countries At Risk

Some countries are even more at risk. The Philippines population is about 115 million, with 48 million workers. About 1.8 million work in call centres, and over 2 million are overseas workers. So at least 10% of jobs in the Philippines are at risk. These industries are critical to the Philippines' exports and standard of living. The fees international migrants paid to banks were nearly as high as US foreign aid in 2023. [38] If you live in the same country as your family, you don't usually have to pay a fee when you send them money. International migrants face a harsher reality: they incur substantial transaction costs when supporting their families back home. Globally, the average fee in 2023 was 6.3%. The estimate is that over $800 billion is transferred from migrants back to their families at home. What happens to poorer countries when it becomes cheaper to deploy AI systems than to pay migrants? The USA has over 2.8 million workers in call centres and over 23 million in public service. What will happen to these service workers? What will be the purpose of public servants? Will people in the education and healthcare sectors be replaced? It is hard to predict, as AI and robotics will enable them to do more, allowing for more productive time in these areas, which is likely beneficial for society.

6.8 Contrast between Countries

The impact of AI on jobs can differ significantly from country to country. Developed countries with advanced technological infrastructure and a skilled workforce are more likely to reap the benefits of AI advancements. These countries can leverage AI to automate routine tasks, enhance productivity, and create new job opportunities in emerging fields such as data science, machine learning, and AI research.

On the other hand, developing countries with limited access to technology and a less skilled workforce may face challenges in adapting to AI-driven changes. They might experience a temporary disruption in certain industries, particularly those heavily reliant on manual labour. However, AI should bring them prosperity and improved health and education, and with proper investment in infrastructure, these countries can also benefit from the AI revolution in the long run. The worst thing they can do is invest in yesterday's conventional infrastructure. Fossil fuels and labour-intensive industries will not be their future. Tourism, education, and nature restoration will drive their economies. Yet we see many heading down the path of old industries.

6.9 140% Increased Productivity

There are two views of AI for service jobs. One is that service jobs will disappear, reducing costs. The other benefit is increased productivity. Both can be true, but we need to look at specific jobs and specific countries to see which one wins out in each sector. Ark Invest states in their Big Ideas [17], [39] that the augmentation leads to a 4-fold improvement in productivity by 2030. An example of productivity improvement is the use of video conferencing. Video conferencing has been around for nearly a century, but it wasn't until 2009 that we saw the rise of Bluejean, Skype, Cu-SeeMe, and, more recently, Zoom, Teams, and Meet. COVID gave the impetus to a whole new industry and changed meetings forever.[17] Here is their logic.

- Humans with appropriate AI will lead to improved productivity. I.e. augmentation.
- The value of the knowledge of humans plus AI will increase by about 15% per year.
- A knowledge worker today earns between $25,000 and $30,000 (global average). The OECD average is $60,000.
- Wage increase is about 2.6% per year.
- By 2030, they would make $36,000 in wages.
- The productivity increase of a worker plus AI means the value of that worker would be $86,000 total.
- $50,000 of this value is from AI augmentation.
- Overall, AI provides 140% improvement in productivity per knowledge worker.

In 2023, Ark stated that spending $41 trillion on AI could increase labour productivity by over $200 trillion, dwarfing the $32 trillion in knowledge worker salaries and rivalling current projections of global GDP in 2030.

6.10 Conclusions of Service Jobs

In the chapter on AI Overview, we looked at the technology and history of AI. This chapter explores the impact of AI on service jobs and the next chapter will look at specific industries. Broad generalisations won't provide where we will see progress over the next decade versus the long-term outlook. AI will reduce jobs in most sectors. Some, such as health, aged care, science or education, may not. We can see AI will severely impact countries reliant on the services industry as those functions go into the AI world. However, jobs won't return to those countries that outsourced them. The jobs will be gone, and forever. Going forward, there will be no point to offshore service jobs . Some are already predicting that soon some entrepreneurs will become millionaires or billionaires without any staff at all. With four out of five jobs in Western countries being service or white-collar jobs, AI will initially increase productivity in most sectors and have a profound impact. The opposing two views of AI for service jobs could both be correct. In the short term, increased productivity will avoid job losses. In the mid to longer term, service jobs will disappear, reducing costs. Both can be true, and we need to look at specific jobs and specific countries to see which one wins out in each sector.

What did we learn from AI Disrupting Service Jobs:

- Individual tasks are already being replaced by AI.
- In the short term, additional productivity could lead to 140% productivity over the next decade.
- Agentic AI will start to do multiple tasks. Jobs will then be lost.
- The benefits in productivity growth will mask potential future job losses.
- Jobs that disappear from outsourced economies, such as offshore call centres, won't come back – they'll simply be gone.
- With four in five Western jobs being service jobs, this disruption is broader than any previous wave of automation.

---

7 AI for Service Job Types

Impact of AI on Service Job Types:

- Review different types of service jobs.
- What is the futures for those service professions?
- Which services will be most impacted soonest?
- Software development, research, and content creation – the white-collar jobs are already being reshaped.
- Call centres, IT helpdesks, and education – the lessons from early AI deployments, including what went wrong.
- Beyond the office – AI for flood monitoring, drones, and city management

There are a few jobs safe from AI and robotics, [44] says The Economic Times of India, and these will be jobs that benefit from AI augmentation. This list is far from exhaustive and is selected because AI is already making inroads into either job loss from automation or job augmentation. We see it now every day. I went to a surgeon, and he stated that he used AI to record his patient records. Did I give my permission? The voice AI listened to the conversation, summarised it, added notes to my patient file, and scheduled the next appointment via SMS and the appointment app. Previously, the surgeon would have had the appointment, spent ten minutes on a Dictaphone, and sent the tape to an offshore transcriber who would have typed up the records. Other employees would have organised follow-up actions. This use of AI (audio, LLM, agentic AI) is trivial, yet a symptom of much larger disruptions.

7.1 Contrast Between Industries

Some sectors are more susceptible to AI and job disruption than others. There will be huge impacts in some industries, and little in others.

- Technology. AI has the potential to create new job opportunities, particularly in roles such as AI engineers, data analysts, and cybersecurity specialists.
- Healthcare. AI won't reduce the number of doctors or nurses, but could make them more productive and help them see more patients. See the Health chapter for other areas of AI improvements.
- Education. See the later section. It is unlikely there will be any reduction in staff, although in the longer term, we have to challenge the notion that we educate young people for jobs when there are no jobs for them after graduation.
- Financial services. The industry can benefit from AI-powered automation to improve efficiency and customer experience, while still requiring human expertise in areas such as risk management and strategic decision-making. Bloomberg reported that over 200,000 jobs are likely to be lost in the banking sector. [40] They do not define exactly what those jobs are. E.g. Bank teller numbers in the USA are still over 340,000, but have decreased 30% since 2010 as digital banking, mobile apps and automation have replaced them. AI is likely to accelerate the end of services, such as tellers and banks, which seem to be focusing on improving their face-to-face customer relationships.
- Manufacturing – most unskilled manufacturing jobs will go. The robotic chapter will explore some of the other implications.
- The hospitality industry will experience job displacement due to increased automation.

Note that there will be new roles in these sectors, in maintenance and programming of AI-driven systems, although within 2 decades, there will not be a job that a machine, a humanoid robot, cannot do.

7.2 AI for Software Development

As to be expected, software is one area of the service sector that has embraced AI to improve productivity. In the section on AI software itself, we looked at the complexity of software development. Not just the software itself, but all the supporting systems from product design to deployment and maintenance.

Coding tools, such as AICodex from OpenAI [41] provide faster development and better software. This AI system translates natural language to code. Another product, Tabnine, claims that they have generated 1 to 2% of all software code over the past 6 years, based on their 100,000 users, and touch 30%-50% of all code. According to independent studies, they gain 20% to 25% improvement in software quality. [42] The ROI at $39 per user per month has yielded $50,000 to $60,000 in monthly productivity gains. Improvements in code writing, code review, documentation, and quality improvement with lower bugs. In contrast with Microsoft's GitHub Copilot, which retains customer data for up to 28 days, Tabnine does not collect, store, share, or use customer code to train Tabnine's models. Tabnine also offers three modes of installation: on-premises and VPC, where code stays within companies' networks only.

One particular issue now raised is the future role of programmers. Today, AI Codex systems can develop better code than junior software developers. Why hire a junior at a $60,000 salary when, for a license cost of $100 per month or less, you can implement AI to write code, test, develop, deploy and document for 90% of the existing tasks. In most software teams, more developers do not translate into fewer bugs and software systems delivered on time.

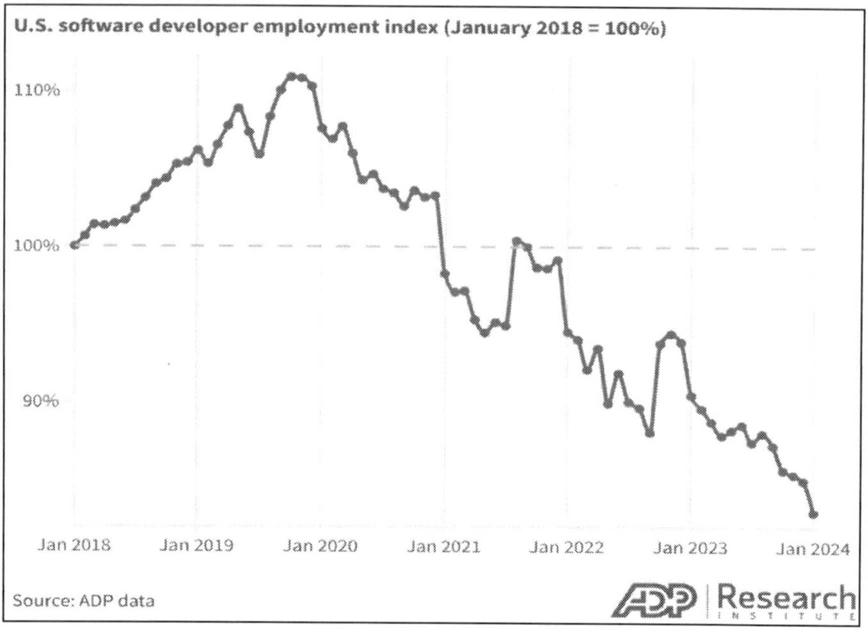

Figure 16 Software Development Job Postings on Indeed in the United States from ADP, using data from Federal Reserve Bank of St Louis.

Figure 16 shows the impact of AI reflected in vacancies for software developers. Vacancies in the USA spiked during COVID but appear to have flattened off from 2020 at a baseline of 100, peaked at over 230% in 2022, and are now down to less than 66%. [43] In contrast, the total labour workforce jobs rose over 150%, and are still currently at 120% of seasonally adjusted long-term levels. Software developer salaries have also lagged behind those of total U.S. workers. Between January 2018 and January 2024, the median base pay for developers grew by 24 per cent while pay growth for total U.S. workers grew 30 per cent. Still, developers remain well compensated, earning $130,000 pa, more than two times the U.S. median salary of $53,000.

Software managers now have the option to purchase AI software that improves the productivity of existing staff. Rather than hire more staff, the company is choosing to use AI tools to speed up development cycles, generate user documentation and provide more robust software, in the style of your existing IT software environment.

What will be the scenario in 5 to 10 years? Will continued improvement further reduce the workforce, or will demand from sectors to deploy AI increase developer jobs?

7.3 AI Tools for Research

The research process has changed little over the past century. Traditionally, a human researcher collects background knowledge, drafts a set of plausible hypotheses to test, constructs an evaluation procedure, collects evidence for the different hypotheses, and finally assesses and communicates their findings. AI has already made huge strides in tasks like writing a paper, or editing a manuscript, or searching for publications, but not the whole process. Tools such as Undermind.ai, Research Rabbit, Jenni, Scite, Elicit, SciSpace/Typeset, Consensus, Connected Papers, Inciteful, Litmaps and products Google LM turn literature search, writing science reviews, and articles little more than 5-minute exercises whereas a science review in 1980's involved trekking to a library, waiting hours, notecards, and four months writing effort.

Extending this process, Lu's team recently announced the AI Scientist, an LLM, which performs the entire process, including generating novel research ideas, writing code, executing experiments, visualising results, describing its findings by writing a complete scientific paper, and then running a simulated review process for evaluation. [44] Their paper was used in Machine Learning, but it applies to other disciplines.

A recent announcement from Microsoft at Microsoft Build 2025 was a similar product. MS Discovery [45] is an AI system that generates ideas, simulates results, and learns over time. They demonstrated how they found a new cooling agent that can be used as an immersion coolant to keep computer system boards cool without the need for fans. The chemical is actually new and, more importantly, does not contain forever chemicals (PFAS). Microsoft says this AI system will have applications in drug development, semiconductors, or new materials. Is science over, or will this open-source product enable science to advance in ways we have not seen?

7.4 AI for Call Centre Helpdesk

AI is already making a major impact in call centres. As AI technology improves, it becomes an easy business decision to hire 10 AI software engineers instead of 200 call centre representatives. The call centre business is one already beginning to transition from people to AI.

Changes are underway. For example, six months ago I had to cancel a house insurance policy as I had changed providers. I went online, and there was no option to cancel and I had to use the chat form, which then led to a phone call with a real person. They asked for my date of birth, the policy number, and my contact details, then cancelled. It took at least 20 minutes for the whole process. Just 4 months later, I had to cancel another policy. This time, the cancellation request took me to a chat window. Again, it asked for my date of birth and policy number, then sent an automated SMS message to confirm my intent. It took less than 3 minutes. That large insurance company would have had millions of policies. A core process is handling cancellations – let's assume 10% per year. That would require dozens or hundreds of call centre employees just doing cancellations. Those jobs are gone. Forever.

Job losses from the use of AI are already underway. E.g., British Telecom (BT) announced plans in 2023 to cut 52,000 jobs by 2030. Their CEO recently said such proposals "did not reflect the full potential of AI."[46] There is a salutary lesson with the Swedish Fintech company Klarna. Six months ago they claimed their OpenAI-powered AI assistant was carrying out the work of 700 full-time customer service agents but with faster response, much better success rates and improved closing of cases. Over that period, staff numbers dropped from 5,500 to 3,000. In June 2025, they backpedalled. Their CEO admitted that relying too heavily on AI led to poor customer service and falling satisfaction levels. With complaints rising and users frustrated, Klarna is now rehiring some humans to restore service quality. The company's bold AI-first strategy didn't deliver as expected. This real-world example shows the risks of replacing people with tech too soon, and why the human touch still matters in customer support.[47]

This field is moving at such a rapid rate, and at many companies, thousands of call centre roles could be replaced by a few engineering roles, as seen with

British Telecom above. OpenAI has released products and the price of their AI audio solution. The price has fallen to $0.06 per minute of audio input and $0.24 per minute of audio output. This means the cost of the AI and systems for a 5-minute call with the Real-Time API is $1.30, or $15.60 per hour, which is the labour cost of an offshore real-person agent today. AI is disrupting the workflow for both customers and agents. If you think about a call centre flow, it used to be multiple steps.

Waited on hold for 30 minutes, listening to the company spruiking its products and services, until an agent answered. [Audio] → human → response. Repeat until the solution, or in many cases, put on hold while the customer service agent escalates the problem to their supervisor. Customers became angry as they were shuffled from one support agent to another or escalated to someone who could assist, supposedly. Everyone has horror stories of help desk interactions. Earlier versions of AI improved things somewhat, but the process was still somewhat convoluted.

[Audio Input] --> [Automatic Speech Recognition] → Transcribed Text] → [Text Model for Inference/Reasoning] → Text Output] → [Text-to-Speech Model] → Audio Output]

With the latest release of LLM models and integration using APIs it is just this.

[Audio Input] → [Chat Completions API] → [Audio Output]

Under the hood, the Realtime API creates a persistent WebSocket connection to exchange messages with GPT-4o. The API supports function calling opens in a new window), which makes it possible for voice assistants to respond to user requests by triggering actions or pulling in new context. With multiple conversations occurring simultaneously, any that require human interaction can be escalated. The process can scale from one to thousands. Every AI agent has the same intelligence. When a new problem is identified and solved, the new edge case is added to the system's knowledge, allowing the next time that problem is encountered, it can be solved without human intervention. Expertise and solutions are fully captured. Scalable, reliable, audited, self-learning. These solutions are still brand new, and, like Klarna, some will rush to deploy them and get burned. Business systems are not just technology, and they take time to plan and deploy. Companies need to spend time planning and trialling their system with their customers. As mentioned earlier, these systems are the dumbest and slowest they will ever be.

7.5 AI for IT Help Desk

IT Help desk systems have been around since computers were invented and are hated by most employees (and employers!) Most help desk software companies, such as SysAid, have integrated LLMs into their help desk systems already. The benefits generally include faster response times, 24/7 availability, process automation, more than just machine learning, scalability, support for multiple languages, and data-driven capabilities. SysAid says that after two weeks, a typical deployment results in IT saving 10 to 15 minutes per ticket, closing 75% more tickets the same day, and improving customer satisfaction. Although AI has only been available for less than three years, companies have experienced reduced staff turnover and improved customer service. Solutions can range from simple password resets to comprehensive solutions. Some are already analysing screenshots when a software bug is detected. They identify the bug in the software, create a fix, and can automate the deployment. We are in the early stages of using AI, but the ability to have instant voice, chat, or email IT support would greatly assist most users, especially in smaller and mid-sized organisations such as schools and small businesses, where IT Help Desk support is often scarce.

Network companies such as Samsara have deployed AI in their network backbone. Samsara is a company that helps businesses track and monitor assets, such as vehicle fleets and equipment, and utilises LLMs to simplify the customer experience and enhance productivity. They connect over 20,000 businesses and collect trillions of data across vehicles, sites and operations. With DHL, they cut accident rates by 50% and halved staff turnover.

The IT helpdesk area is an example of how change management is much easier than in some other service. IT staff are computer literate, they understand the technology and are used to constant change. Dealing with staff who have low IT skill levels can be painful for staff who sometimes lack emotional intelligence. Trained experienced support staff are limited. The professional services business Kyndryl, spun out from IBM in 2021, focuses on deploying agentic AI to its customers. They point out that the provision of AI IT support systems is very acceptable. The change culture in IT services is amenable to deploying new "untested" systems; there is a very good return on investment for IT improvements; and successful implementation provides a basis to extend to other business processes.

7.6 AI for Education

Education currently is as much about learning how to be bored for long periods of time, preparing for work where workers get bored out of their minds and hate their jobs. AI is making significant strides in the field of education. However, foundational skills remain at the core of learning, and AI should serve as a complement rather than a replacement. While AI can undoubtedly enhance education, it's not a shortcut. Reading, writing, and critical thinking remain vital for a well-rounded education. AI should be thoughtfully integrated to support and amplify these skills. UNESCO [48] suggests application in educational contexts needs to be guided by the core principles of inclusion and equity.

Over-reliance on AI in education may prove counterproductive, potentially impeding the cognitive processes crucial for intellectual and creative development. In education, the goal isn't simply to accumulate information but to develop analytical, synthesising, and application skills.

That's what they said when slide rules became the mathematics aid. It is what was said when calculators became affordable. The challenge is now that all of the written words since the printing press was invented are now available, via the internet, available on your wrist and now presented to the user in a coherent stream.

7.7 AI for Authoring Content

Over the past century, the cost of authoring written content has been relatively constant in real terms. [17] Over the past two years, as LLM writing quality has improved, the cost has collapsed from $300 per 1,000 words to under 4 cents. In any language you want.

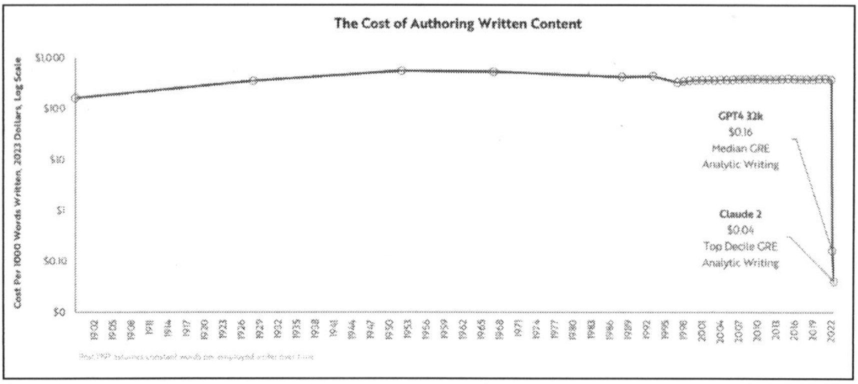

Figure 17 Cost of authoring Written Content by Ark Invest 2024

Figure 17 shows the cost for writing has fallen over 7,500 times. Authoring content for marketing material, e-books, training materials and documentation for software, help manuals – documents are everywhere. From the bedside light with a safety warning to Ikea assembly instructions, words written by someone is everywhere. LLMs can provide those at a fraction of the cost and, importantly, the time.

7.8 AI for Literature

AI LLMs are powerful in literature. I use breaking the writer's block as a simple strategy when approaching a new chapter and don't have a solid framework. So, for this small section, I asked Gemini to "provide 5 use cases for AI in literature", and in 5 seconds, I had some pointers. I revised the output and added to it. When I asked Grok4, I had different uses cases. The benefits of AI in writing include:

- Supercharged Research. AI can sift through vast amounts of literary works, helping researchers identify influences, recurring themes, and connections across vast datasets much faster than ever before. There are now specific AI tools that can search for all recent publications and compile a master 's-level literature review in seconds.
- AI-powered Analysis. AI can analyse stylistic elements like word choice, sentence structure, and narrative flow to distinguish genres, identify authorship, and even predict reader reception.

- Personalised Storytelling. AI can recommend books based on your reading preferences, or even create interactive stories that adapt to your choices, transforming the reading experience.
- Breaking Writer's Block. AI tools can help writers brainstorm ideas, generate descriptions, or even translate languages, providing a springboard for creativity and overcoming roadblocks.
- Accessibility for All. AI can transform text into audio, summarise complex passages, or translate languages in real time, making literature more accessible to people with visual impairments or language barriers. Amazon and other publishers have integrated AI into the whole book publishing process. You can upload a document, and AI proofreads the book, creates cover pages, creates text for the back cover and for online shops, creates books, audio and e-books and sends to multiple publishers.

I then submitted the above points to Gemini and asked if they were AI-generated. Gemini responded:

> *In short: The text is well-written and plausible as human-generated content. While AI could have produced it, there aren't strong indicators that it definitely did. It's more likely a human wrote it, possibly drawing on information they've read about AI in publishing.*

An increasing number of articles are generated by AI, to the extent that AI-generated content is ubiquitous. AI tends to use repeated common words: "showcasing, underscore, navigation, meticulous, pivotal" and others. They are often flowery, and lack a sense of individual writing. The lack of individualisation is not really a problem. AI is a tool. ChatGPT writes with us, nor for us. They can be a virtual assistant, although originality and authorship can be lost. There are numerous benefits, and individuals who struggle with English as a second language utilise AI to enhance their communication. LLMs can generate words but they may not generate wisdom. Given the vast investment in computers and software, AI is improving rapidly, and we are likely to find more and more use of AI. A consequence is that there should be less poorly written material. There is no longer any excuse for flawed writing.

While the above may seem positive, there are those who are hostile to the whole notion of AI in literature and say that AI takes away from being human.

It is a double-edged sword. Another issue is where the LLM is getting its source data from? AI slop is now ubiquitous. Read any social media post or news article and ask yourself, where did it come from? A writer or an LLM? Some LLMs, in their training, will use information from websites and social media, and the old adage of "garbage in, garbage out" is obvious.

7.9 AI for Graphic and Creative Arts

AI has a role for artists, whether they are in the performing arts, in painting or in music. There is a heated debate about the role of AI but already it is transforming the industry and it is hard to predict the future. Looking ahead:

- Graphics: AI in graphic and design tools is well on its way to full AI integration. Most graphic design packages have transitioned to AI-based tools to help designers increase productivity, with many software solutions now incorporating AI as part of their offerings. [49] Marketing collateral, flyers, product information and website development can now be done with a large number of AI enabled software packages. Every modern graphics software program, including Photoshop, Canva, Microsoft Office, and Google Docs, has AI agents to create graphics and media assets. Every week, there are new AI software products released. A recent product can take an AI-generated meeting summary and transform it into a structured presentation, complete with graphics and relevant information.
- Gig Economy Disruption: Upwork has over 18 million freelancers and 841,000 active clients (spending more than $5,000 per year). Fiverr has 5.5 million active buyers and 380,000 sellers. The industry is large, with over 64 million Americans performing freelance work in 2024, representing 38% of the US workforce. Most would be using AI tools, whatever the gig.
- Inspiration and Brainstorming: AI can help artists overcome creative block and generate new ideas. For example, an artist working on a portrait could use an AI tool to create variations on facial features, hairstyles, and backgrounds, sparking new visual directions for their artwork.
- Art Restoration and Preservation: AI can be used to analyse and restore damaged artworks. By analysing patterns and details in

undamaged areas, AI can fill in missing pieces or remove unwanted alterations, helping to preserve cultural heritage.

- Personalised Art and Design: AI can create custom artwork or design elements based on user preferences.
- Interior Design: Designers now use AI to generate furniture arrangements and colour palettes tailored to a client's taste.
- Interactive and Immersive Art Experiences: AI can be used to create dynamic and responsive art installations. For example, an AI-powered sculpture could react to a viewer's movements or emotions, creating a more engaging and personalised art experience.
- Music Composition and Performance: AI can assist musicians in composing music and even perform alongside them. An AI program could generate new melodies and harmonies based on a musician's style or create backing tracks that adapt to the musician's improvisation.

At what stage and what level of AI is acceptable?

7.10 AI for Movies

The Writers Guild of America went on strike in 2023, partly to address the impact of streaming but also to prevent the use of AI (such as ChatGPT) in any production. Since 2023, AI tools for graphics and video have surged. New free AI engines can create multi-minute-length videos, and text-to-image and text-to-video are now part of most LLM tools.

- Special Effects (VFX): AI can automate tedious tasks in VFX creation, like removing green screens or rotoscoping (separating a foreground object from the background). This frees up artists to focus on more creative aspects. E.g. In "Avengers: Endgame" (2019), AI was utilised to remove the bulky facial capture suits worn by the actors during filming, resulting in more realistic and expressive characters, such as Thanos.
- Character Animation: AI can be used to create realistic and nuanced movements for 3D characters. This can involve learning from real-life motion capture data or even generating new animations based on a script or storyboard. E.g. In "The Lion King" (2019 remake), AI was used

to animate the realistic fur and muscle movements of the lion characters, creating a level of detail that wouldn't have been possible with traditional techniques alone.

- Image/Video Editing: AI can automate repetitive editing tasks like colour grading, noise reduction, and even content-aware scaling (enlarging an image without losing quality). E.g. Adobe's Premiere Pro, a popular video editing software, uses AI to analyse footage and suggest colour correction presets, saving editors time and effort.
- Storyboarding and Pre-Visualisation: AI can generate storyboards or previz (preliminary visualisations) based on a script or concept art. This allows filmmakers to plan and visualise scenes before filming begins, helping to identify potential problems and ensure a smoother production process. E.g. the movie "Alita: Battle Angel" (2019) used AI to create previous sequences of complex action scenes, allowing the director to experiment with different camera angles and choreography before committing to expensive live-action filming.
- Content Personalisation: AI can personalise movie trailers or even tailor the visual style of a film based on a viewer's demographics or preferences. This is still in its early stages, but it has the potential to create a more engaging viewing experience. E.g. some streaming services use AI to recommend movies based on a user's watch history. In the future, AI might be able to subtly adjust a movie's colour palette or visual effects to better suit a viewer's tastes.

7.11 AI for Sign Language

Google is working on SignGemma, which converts sign language into spoken words, and will release it this year. The days of braille and sign language are passing. AI will be a technology for those without language.

7.12 AI for Telemarketers

Tired of those pesky calls trying to sell all things under the roof? Solar, real estate, vehicles or cryptocurrency? Telemarketers are likely to become an endangered species. But there's no need to be overly joyous as the human voice is likely to be replaced by more annoying personalised and computerised calls.

When you know the voice on the other end is a machine, will you be less tolerant and more aggressive?

7.13 AI for Flood Monitoring

Recently, Australian researchers discovered that artificial intelligence-driven processing of data derived from existing 5G mobile networks, in conjunction with a digital twin – a virtual replica that mirrors its real-world counterpart – can deliver unprecedented accuracy in flood monitoring and prediction.[50] No new equipment is needed. No sensors, no new towers. They claim to predict, with 90% accuracy or higher, which roads and houses are likely to be inundated within the next 24 hours. Within a few seconds, down to a 5m x 5m grid and 0.1m height. Existing systems cannot accurately predict flash flooding, whereas the AI/5G has proven it can. The world will wake up to the fact that we will have higher intensity storms. Some think the floods in Valencia, Spain, in 2024, with over 230 deaths, or the floods in Texas in 2025 that took over 200 lives, are rare events. Unfortunately, climate change will increase the likelihood of such disasters, but AI can provide a warning system.

7.14 AI for City Management

China is leading in traffic management with projects such as Hangzhou's "City Brain", an AI platform managing traffic flows and emergency responses in real time. [51] Developed with tech giant Alibaba, City Brain system collects billions of data points from traffic cameras, "Internet of Things" sensors, police reports, toll stations and municipal records, all in real time. The goal is to centralise and algorithmically manage the city's operations with a level of efficiency that only AI can provide. The City Brain was rolled out to combat congestion in 2018 to a limited area, but was extended over the years. It helped transform Hangzhou's gridlocked roads. In 2020, the provincial capital south-west of Shanghai was the second-most congested city in China. Within a year of AI-powered traffic light optimisation and policing decisions, it had dropped to 34th. The system didn't just ease traffic - it slashed accident rates and accelerated emergency vehicle response times. Other cities are adopting the technology. City Brain systems have been promoted as making cities "smarter" and improving their residents' quality of life but the systems have also faced criticism for issues relating to privacy, cost, and their use of surveillance.

City Brain 3.0 (2025) goes further into city governance, not just roads but water, geological surveys and city management. [52]. DeepSeek AI virtual assistants provide services, including Jingxiao'ai, a virtual police officer who offers 24/7 legal and administrative assistance, quickly guiding residents through tasks such as household registration applications. Hanghaomeng, China's first AI expert in mental health, has already served 1.8 million users, providing online consultations for sleep disorders. Yibao'er, an AI system developed by the city's healthcare bureau, streamlines medical insurance inquiries and transactions.

7.15 Which SectorsAre Safe?

As we have scanned across industry sectors, the question has arisen: which jobs are safe? Some jobs appear safe today, especially management jobs, where decisions are made in deploying AI systems. Any HAS1 or HAS2-scale jobs or augmentations are probably safe, but jobs that are HAS4 or HAS5 are at risk.

- Positions in AI that develop and deploy new AI systems, data science, machine learning, and ethical AI oversight are jobs that will have a longer-term career path. General IT support jobs are at risk.
- Traditional skilled manual roles—such as plumbers, electricians, and decorators are probably fine for the next decade or two. Still, as we discuss in the next chapter, humanoid robots will become more able over time. When humanoid robots are widely available, they will only need six months of training to end that trade. A skilled tradesperson could wear a video camera for a few months, and AI would consume the video, providing that expertise to a robot. The only constraint is the dexterity of the humanoid robot. These trades have nothing specific or unique that cannot be replicated. Moreover, the humanoid robot will be aware of all the regulations.
- Medical jobs – AI is more likely to support medical people.
- Education and training jobs are already at risk.
- Science and technology jobs are not at risk, but augmentation will become a bigger part of the role.

7.16 Conclusion - Service Sectors

AI is predicted to impact 40% of jobs overall or higher. Jobs in advanced economies may benefit from AI integration in the short term. It is less clear about the longer term. Low-income countries are not immediately at risk. Sectors particularly at risk include:

- Software and computers.
- Authoring.
- Graphics.
- Public services, including City Management, compliance, natural disaster management and planning.
- Finance services.

Estimates for productivity increases range from 10% to 140%. Organisations and industry sectors that have adopted AI have seen improved productivity. At the same time, some industries have adopted AI to reduce costs and have discovered that their goal for reducing labour is being met or surpassed. Jobs at immediate high risk include those that were outsourced to call centres 20 years ago. Those call centres have streamlined and consolidated support for Western businesses, but are now being disrupted themselves. Sectors that will reduce employees include.

- Finance sector.
- Accounting sector.
- Call centres for health, insurance, banking, travel and public services.

It is easy to accept the introduction of AI when productivity increases, customer support improves, and bureaucracy is reduced. It is more concerning when we see job losses and career pathways stymied. What jobs will be available for our children and grandchildren? Be a plumber, as Geoffrey Hinton says? We have all had experience with new technology introduction, and no matter what the advocates say, new technology is never as easy as it is made out to be. The major concern for all is what happens to those who lose their job? If people need work for self-esteem, and that job is no longer there, what is their purpose in life? We will discuss this in a subsequent section.

Conclusions from AI and Service Job Sectors :

- Many tasks have already been AI'd.
- AI will augment people in some sectors in the short term.
- Job loss is happening now.
- Job loss is increasing.
- It is hard to identify any job sector that will not be impacted.
- Klarna's AI-first stumble shows the risk of moving too fast – but also that the technology genuinely works

8 Robotics & AI

Introduction to Humanoid Robotics:

- Why Now, Why Here?
- Technology of humanoid robots.
- R&D status and early developmental use.
- Manufacturing challenges and autocatalysis.
- Compelling economics - the cost of labour by humans or machines.
- Can all jobs be done by machines by 2045? Will they?

Is laziness the default for all human beings? Is it hard-wired? Why do we minimise labour? It is all about energy. Calories were scarce for virtually all of our species' existence up until the last few decades. Expend 10 calories too much each day, and you die. Laziness is built into our very genetic code. How many problems in the world are born of neglect - from litter to dilapidated streets and buildings to the dirty dishes in the sink and unfolded laundry in the hamper - come down to energy? Robots will not be lazy. They will enable a world in which marginal labour costs approach $0.50 per hour. It is not just about making a post-scarcity world, but one with super-abundance. What would that look like?

Robots disconnect physical labour from output. The economic output of a company or a country is principally determined by the labour times the productivity of that labour, and accounts for 50% or more of most countries' GDP, as we saw mentioned briefly previously. Robots address both supply and productivity. Labour becomes unlimited. Productivity improves over time. Humans generally do not become more efficient. Consider the "productivity of Australia" [53], which grew from an index of 80 in 2000 up to 110 in 2019 but has fallen back to under 100 by 2025. [54] Why? Evidence suggests that at least part of the decline in economic dynamism is due to reduced competition, but there are other factors as well. Part of that increase over the past 20 years is that automation has transformed large industries. As far back as the Industrial Revolution, machines have taken over tasks. While we might not recognise a washing machine as a robot, it is a specialised robot for washing clothes. Time for laundry fell from 14 hours per week to less than 2 hours, an 87% reduction. Robots have reduced the time to manufacture a car from 12 hours to 2 hours (88%), and Kiva Robots have slashed the time to ship from an Amazon warehouse from 70 minutes to less than 12 minutes (78%) [39]. Later in this chapter, we will discuss how broader adoption of general-purpose humanoid robots will disrupt this productivity measure.

Both RethinkX [55] and Ark Invest's models of disruption [17] detail how AI and robots would lead to a reduction in labour across all sectors. Both call for some form of universal income to avoid economic hardship. AI is the brains of robots and a key factor in the rapid increase. Some call humanoid robots "Cobots" – robots that are cooperative with humans in the workplace. We will focus mostly on robots in human form, not "industrial robots." While industrial robots are already widespread in factories, humanoid robots will be the most disruptive.

8.1 3 Perspectives on Robots

The rise of the mass market for humanoid robots will create conflicts among technology, economics, and politics, with less than 5 years before we get swamped. For Tracey, it might be 5 years. For Ada it will be about a decade. This global technology will disrupt economies and lives. We could ignore humanoid robots if all they did was work in large factories alongside industrial robots. But humanoid robots are likely not only to expand to use in all sectors of the economy but also to overwhelm it. There is no reason to doubt those who predict there will be more humanoid robots than humans. The only uncertainty is how many times more. We need to be ahead of this rollout, or this will be a fait accompli without addressing the many socio-political challenges raised. Let's assess the future from these three perspectives, as we like to do for most innovation, to tease out why we are for or against new technology.

- Technology: The technology for industrial robots is mature and still improving, but the technology for humanoid robots is still at the R&D phase. Some are at the "prototype pilot manufacturing" phase. Companies are trying to make the product functional and manufacturable. There are two components. The physical and mechanical component of the humanoid robot, and secondly, the brain or the software and control components. What engineering challenges are left to solve? Will humanoid robots be in widespread use by the end of 2026? How rapidly can production scale?
- Economics: How compelling are the changes? Will these accelerate or slow down adoption? Why would businesses, individuals, or society adopt humanoid robots?
- Political and belief system challenges: Will some agents support the changes? What are the implications for individuals, businesses and society? What will the scenarios be in five years and 20 years? Many shudder at the thought that humanoid robots will be "living" with us, in close proximity. In our kitchens, living rooms and bedrooms. Is it science fiction? It is probably closer than we could imagine.

8.1.1 Why Now?

The future of robots is not what we see today. The advent of humanoid-form robots will create a tsunami of change. We have many types of robots, although we may not recognise them as such. Domestic robots such as washing machines or other labour-saving devices are single-task and generally dumb. Manufacturing robots are typically fixed and designed to perform one or many tasks at a time. We are familiar with auto assembly lines where cars are churned out, or soft drink factories where millions of cans come off the line. During the bottling process, various workstations may perform many tasks, including washing, drying, filling, affixing labels, screwing on lids, and placing bottles into cartons at breakneck speed.

These assembly lines, popularised by Henry Ford's car manufacturing, have evolved over the years, and tasks humans once performed are now performed by machines. The deployment of industrial manufacturing robots is booming. These industrial robots are fixed in place. The surrounding factory facilities are modified to suit the robot. If they are dangerous, they have fences to keep people out. The robot might only cost $20,000, but the fitting of the factory around them could cost $100,000. Robots range in size from micron-sized robots used in chip manufacturing and electronics to huge robots in mining or manufacturing. Many smaller businesses cannot afford that level of capital investment or cannot justify the productivity gain. Instead they employ people to do the tasks. Sometimes they invest in a production line capable of producing a range of products. Even in industries we consider highly automated, there are many tasks that are not economically feasible to deploy a robot.

The significant change is that robots will be available in a human form. Industrial robots are here to stay and continue to be widely deployed because they save labour, provide high throughput and offer a good return on investment. What they are not is flexible! We will explore how they have progressed over the past 2 decades.

On the other hand, humanoid robots are highly flexible and will work in our human environment, often alongside humans. With no need to modify the environment, these robots can perform multiple tasks across multiple locations alongside human workers. These new robots are designed to work alongside of humans. They are built to avoid injuring humans either unintentionally or intentionally. Lightweight and not strong enough to injure or overpower a human. (Military robots will have different designs.) They could stack items into boxes in one task, then carry those boxes to another site for storage, and sweep up the floor in another. A domestic humanoid housekeeper robot could do cleaning, ironing, bed-making, gardening, folding clothes, buying groceries, cooking, washing dishes, and driving a vehicle to serve as the children's taxi service.

> *Now, we are on the cusp of a new disruption: physical labour performed by humanoid-form robots. Except this time, we are the horses. The disruption of labour by humanoid robots* (Tony Seba / RethinkX) [56]

Just as internal combustion engines gave automobiles the capability to disrupt horses, a convergence of technologies that together create a "labour engine" is what gives humanoid robots the ability to disrupt human labour. Industrial robots have created some disruption, but they will be a fraction of what is coming with these new humanoid robots.

8.1.2 Tsunami of Robots

Technological convergence could create tectonic macroeconomic shifts more impactful than the first and second industrial revolutions. Robots have already made an impact with industrial robots.

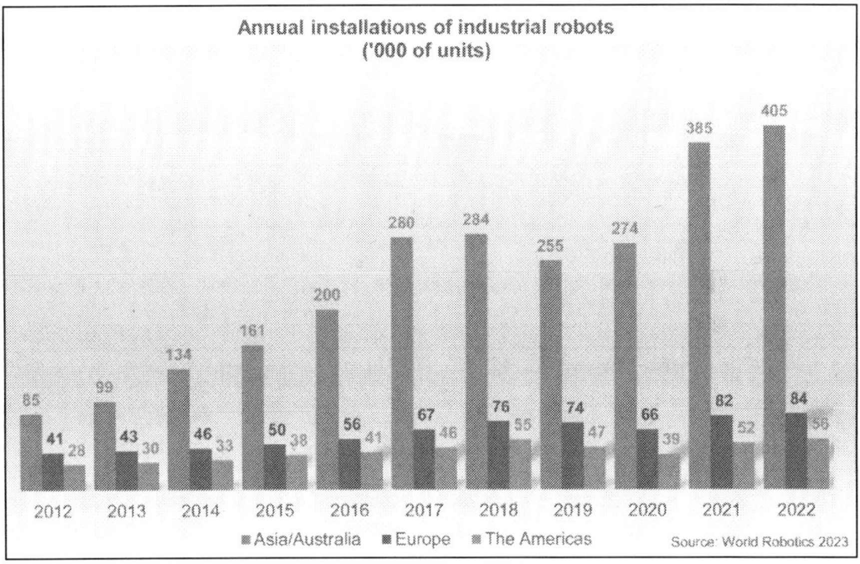

Figure 18 Industrial robots new installations from 2012 to 2022

Figure 18 shows the large rise in industrial robots. Globally, the total number of robots in service exceeds 3.9 million, and that number is rapidly increasing, with over 553,052 industrial robot installations in 2022, representing a 5% year-over-year growth rate Over over the past decade, over 73% of robots were installed in Asia, [57] 15% in Europe (720,000) and 10% in the Americas (495,00 total). China had 1 million in 2021, but 1.5 million in 2022. The five major markets for industrial robots are China, Japan, the United States, the Republic of Korea, and Germany, and combined, those countries have 79% of all robots.

Why have electronic goods (TVs, appliances, phones) become so cheap? Because that industry is a major user of robots. In Asia and Australia, over the past decade, the number of robot installations increased 6-fold, from 85,000 per year to over 400,000, whereas in the Americas and Europe, they have only doubled, from about 40,000 per year. America is a laggard. The current emphasis from the current USA administration on increasing domestic manufacturing is not reflected in more robots.

Additionally, the rate of deployment of robots is rapidly increasing: Amazon deployed over 200,000 over 7 years to 2019, and in the last 2 years, it deployed a further 150,000, and it is the largest business globally deploying robots. Ark Invest (2023) say that robots are at an inflection point. In 2019, there were 1,300 per 10,000 employees in the auto industry, but at Amazon, there were 3,200 per 10,000 employees. The prediction? Industrial robots will continue to scale at 80% per year for the foreseeable future, but a whole new market of humanoid robots will develop, far exceeding the industrial robot market.

> *"Advances in computer vision and deep learning have improved robot performance 33-fold in seven years. Robots are already surpassing human performance by a factor of greater than two and it's unclear where the upper limit will be" Ark Invest 2024* [17]

8.1.3 Experts Underestimate Speed of Change

Experts get it wrong. They got it badly wrong with solar panels, batteries, and vehicles, as RMI [1] shows below. These are physical products, made in factories and only really assembled on site.

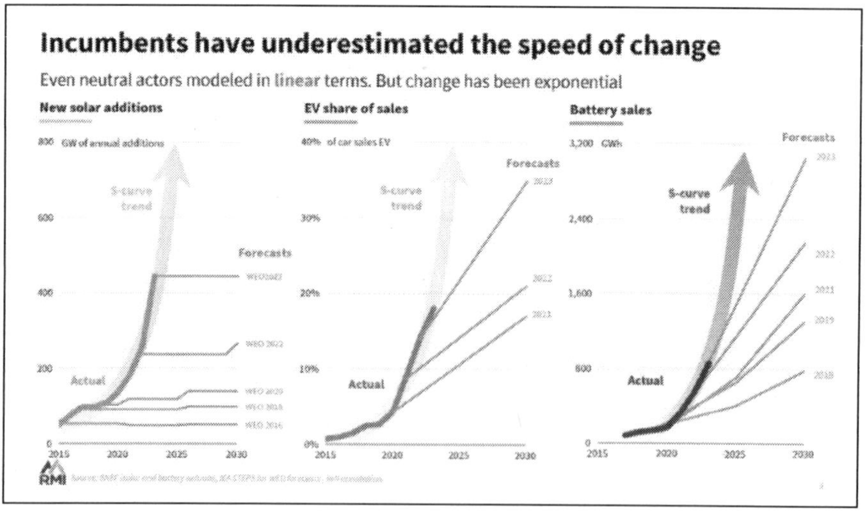

Figure 19 Incumbents underestimate the speed of change

Figure 19 shows incumbents failed to spot the rise of solar, electric vehicles and batteries. The only constraint for AI in the service industry will be the change management required in businesses, and that will be limited by the number of people who can help in the deployment of systems. AI can develop its own code and scale. The barriers to faster deployment is the sales process and internal change management process. Industrial robots will continue to grow as they outperform humans in factories.

Humanoid robots will be different. Hardware and software costs decline according to Wright's Law, but AI is improving at 3 to 4 times each year. Together, they will improve productivity and create new market opportunities for robotics. A production rate of 100 million robots per year, each costing $25,000 is a $2.5 trillion market. Is this unreasonable? If we compare the market to cars, about 80 million cars are produced each year, or to phones, a billion. Ramping up to make 100 million units for something that is much simpler than a car is entirely possible. But that is just in the first 3 years, and we have discussed autocatalysis, where humanoid robots make humanoid robots. There will be no requirement to employ and train a whole workforce. Humanoid robots can work 22 /365 to make new robots.

The ramp-up could be slower or faster than models predict, and the only limiting factor will be the new industry's ability to produce enough. Humanoid robots are already hard at work. Many experts predict that for the next 5 years; most humanoid robots will be in factories and undertaking tasks, yet some robotic companies have said they will focus on the consumer market. What is certain is that by 2030, they will be sufficiently versatile and affordable for deployment in homes. Within a decade, there will be as many robots as humans. Over the past 3 years, the estimated manufacturing cost has decreased from $100,000 to $20,000 or less as the robots move from prototype to pilot to preproduction.

8.2 Humanoid Robot Technology

Even though industrial robots have improved substantially, there is a raft of technologies combining to create a smaller, cheaper robot. To mimic a human, technologies to drive improvements include:

- Sensors: Cameras, tilt sensors, pressure sensors, microphones, accelerometers, etc.
- Computer hardware: In the chips section we looked at the development of AI inference computers by Tesla and Nvidia.

- Artificial intelligence: The robot needs an AI brain to operate in an unstructured environment.
- Electric motors: hydraulic motors are too slow and too costly.
- Precision movement: A body and frame to be able to have the same movements as a human, so the robot can go into all the places an able-bodied person can, without modifying the environment. A humanoid robot has to be able to climb stairs or traverse uneven surfaces.
- Fine but strong hand movements: Have to be able to pick up a pin, a baby, a t-shirt, a cup of coffee or a sack of wheat.
- Batteries: Batteries and charging systems enable independence and power.

Each of these technologies has become dramatically cheaper and more powerful in recent years, setting the stage for a new type of robot. As we will see from the robot announcements from dozens of companies, most challenges have been solved. Improvements and refinements will continue for years. Hardware has reached a point where the humanoid robot can do most things. Improvements are likely incremental, rather than transformational. It took decades for humanoid robots such as those from Boston Dynamics to reach the point where a robotic dog could provide assistance in search and rescue. The improvement of hardware and software is due to improvement in 4 key areas.

8.2.1 Electric Motors and Magnets

A single technology has driven the advance in hardware improvements. Stronger magnets. Humanoid robots need electric motors. Small but strong. Why? Before the 1990s, the power density of hydraulic motors was approximately 100 times greater than that of electric motors. Today, that gap has been reduced by a factor of 10 due to advances in the strength of permanent magnets, such as Neodymium (Nd-Fe-B). [58] Instead of hydraulics, which Atlas from Boston Dynamics used in 2013, designers can use electric motors. Electric motors are lighter, faster, don't leak, are controlled directly with electronics and are easier to manufacture. [59] This is one of those examples of disruption where disruption comes from outside the industry. The inventors of the permanent magnets did not realise they would displace hydraulics, and they almost certainly did not recognise that humanoid robots would soon become a trillion-dollar global business. The change in electric motors has driven disruption in cars, where electric cars have 20 moving parts and are displacing internal combustion engines with an excess of 2,000 parts. Furthermore, the energy efficiency of battery electric cars is over 85%, whereas combustion engine cars are about 18%.

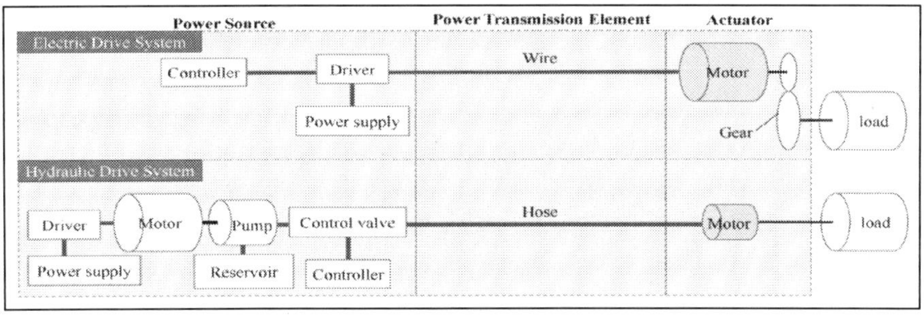

Figure 20 An electric drive system versus a hydraulic drive system.

Figure 20 from Sakama et al shows how smaller and more powerful electric drive now surpass hydraulic drive systems. With supermagnets, motors have become stronger and lighter. Changing from a hydraulic system to an electric drive system leads to simplification. Electric motors have a faster response, fewer parts, smaller size, and are more robust with fewer moving parts, all contributing to lower cost, easier manufacturing, lower servicing and better functionality.

Three years ago, there were no small, powerful electric motors (actuators) suitable for use in humanoid robots. Multiple companies have now developed those actuators, which are fast-response, powerful, precise, and can be manufactured in the billions. Some of the fine motor controls needed for hand manufacturing are still at that early development phase and we can expect continued rapid improvement. A recent video of robots running a half-marathon in China showed how the speed has increased an order of magnitude just in the last year.

8.2.2 Artificial Intelligence

While magnets have been instrumental in "hardware" solutions, the "software brain" of the robot has also made significant strides. AI excels at language and abstract tasks like playing chess and even performing complex surgeries under the watchful supervision of a surgeon, but fails miserably at basic physical reasoning like folding laundry or managing households. Why? Because reasoning about physical interactions and applying abstract skills to dynamic environments are far harder challenges. The human brain – even that of a four-year-old – has been pre-trained for thousands of hours in spatial and physical reasoning. A four-year-old who has been awake for 16,000 hours in their life has about 10 to 15 exabytes of information in their visual cortex. Robots, by contrast, start at zero and operate in brittle, rule-driven systems. Humanoid robots are similar to vehicles in that they involve complex processes. We saw Tesla announce, amid fanfare, in 2016 that they would have self-driving cars in just a few years. To the chagrin of vehicle owners, it has taken a decade. In 2022, Tesla said it would throw out millions of lines of code and move to AI neural nets. The AI needed to train a humanoid robot is ten times that needed to train an autonomous vehicle, claimed Elon Musk on Tesla's 2024 quarterly report [60] and more recently claimed it was 1,000 times harder. With good planning by Tesla, they have some of the worlds most powerful AI training centres and also have years of video of real cars, driving in the real world. Unlike vehicles and roads, which have clearer boundaries and constraints, humanoid robots operate in a diverse and ever-changing, free-form environment. Roads have generalised rules: a human environment is diverse, ever-changing with no rules.

> *I want AI to do my laundry and dishes so I can do art and writing, not for AI to do my art so I can do my laundry and dishes* Joanna Maciejewsks

Previous chapters have more detail about AI advances. Mohapatra [61] details the following three main processes needed for robotics.

- Perception: Convolutional Neural Networks (CNNs) and their successors allow robots to interpret high-dimensional sensory data. Robots need to detect objects, scenes, and depth to understand their surroundings. They have to map their surroundings in 3D. They also need to have touch or something that equates to it. For example, if a drink is spilled, they need a sponge to clean it. Would they need physical touch after they had wiped up the liquid? The bench would still be damp. Did they then need to use a dry cloth after using a sponge?
- Planning and decision-making: Once the environment is mapped, the robot needs to asses and determine its actions. Recurrent neural

networks (RNN/LSTM) facilitate decision-making, such as when to lift its foot while walking, when to turn around and pick up the box that needs to go onto another conveyor belt, or how to pick up the shirt and fold it onto the ironing board before grabbing the iron. Imagine the decision-making challenge of picking up a bottle, unscrewing the lid, and pouring a glass of soda, versus picking up a glass and putting it in the recycling bin. Tasks we could do without conscious thought would be challenging for a humanoid robot. They do not need to relearn that process, and they can share that learning with other robots.

- Actuation/Control: The final step is to control motors. This can be done by low-level controllers (PID loops), but more recent advances use neural nets to learn a mapping from intended motion to actual joint torque. What joints should you move to maintain balance? How much strength is needed so it can pick up a glass? It needs enough to get a grip on the slippery glass, but not too much or they would risk crushing the glass.

State-of-the-art humanoid robotics draws on reinforcement learning to acquire complex behaviour, imitation learning to leverage human skills, model-based control for stability and efficiency, and deep neural networks throughout the perception-to-action loop. Humanoid robot companies' research aims will focus on integrating these components seamlessly. This fusion enables robots to perform increasingly complex tasks. Challenges such as generalisation and safety remain. Tesla's Optimus and Figure AI's humanoid are initially aimed at industrial and commercial tasks (such as moving packages and working on assembly lines) to address labour shortages or perform dangerous work. The AI for this domain will focus on reliability, precision, and integration with existing workflows. The AI for an aged care environment or for a bar waiter will be quite different. The AI model will have some general-purpose neural networks and then add specific role-specific neural networks. We've seen this approach in Tesla's autonomous driving development, where there is a generalised road rule model. It then applies modifications depending on whether the vehicle is being driven in California, Texas, China, or Australia.

The research task has got much easier. Nvidia has released specific computer hardware and software to accelerate the control needed for robots. Their open-source software enables even junior software developers to develop complex systems. Their computer systems are designed from the ground up to support neural net computing from the multi-terraflop datacentre to an onboard humanoid controller. They even have created an environment where school or university students can build their robots. A thriving ecosystem means barriers to entry are low, but the challenge has shifted from building a world-beating robot to producing them at scale.

8.2.3 Batteries

Humans can work in a shift, which may be 40 to 60 hours per week. During the Industrial Revolution, managers realised that working longer was unproductive, and over the intervening years, a productive week's work has generally been settled on 40 to 60 hours. While 10-16-hour workdays were common in 1914, Henry Ford's experiment with an 8-hour workday led to a doubling of profits. With robots, that whole concept of shifts and downtimes and productivity is shattered. Robots will work at the same pace, all the time. And make the same quality, every time. No more Monday morning lemons after a hard weekend of partying.

It is unlikely that auto makers realised they would be disrupted by music. The commoditisation of music-player batteries gave rise to portable computers. Advances in portable batteries led to an EV automotive industry. And a grid battery storage industry. And the end of the fossil fuel era.

Humanoid robots will be able to work 22 hours per day, 365 days a year, allowing for repairs, maintenance and charging time. Having robots tethered to a power cord might be ok in a few work environments, but generally, they will need to be independent and mobile. Batteries enable that flexibility. Energy density and lifetime is increasing, and charging speed decreasing. Good news for humanoid robots.

8.2.4 Fine Motor Control of Hands

The human hand is extraordinary with its fine motor skills. It can be used to work a shovel or play the piano. The human hand has 27 degrees of freedom (DOF). This includes 21 DOFs for the fingers, 3 for the wrist's rotation, and 3 for the wrist's translation. Each finger has 4 DOFs: 3 for extension/flexion and 1 for abduction/adduction, with the thumb having 5 DOFs due to its unique joint structure. The Figure robot has 16 and Tesla's Optimus has 22 DOFs, up from an initial 12. Tesla claimed the robot could play a piano by the end of 2024 but that has been and gone. While Tesla has the most DOFs today, it is unclear how essential that degree of control is needed for most tasks. Some have similar fine motion control, and others have more limited motor skills but can do tasks such as picking up and carrying boxes where such fine motor control is not needed. Eventually expect all humanoid robots to have hands that are equivalent or better than human hands. The fine motion enables the robot to work in any human environment, without having to change the environment. The robotic hand had not had the sensitivity that a human hand has until about 2024, but sure enough, researchers report in June 2025 that they have figured out how to get artificial single-layer skins. [62] Chinese have also developed sensitive hands able to play the piano. [63]

8.2.5 Robotics R&D Status

The status of humanoid robots is ever changing with announcements of advances, seemingly made daily. Morgan Stanley have a useful report [64] Significant activity is in the United States, China, Japan, South Korea, and Europe. Prominent players include:

- USA. Tesla (Optimus), Agility Robotics (Digit), Boston Dynamics (Atlas), Figure AI (Figure 01), Apptronik (Apollo), Sanctuary AI, 1X.
- China. UBTECH Robotics (WalkerX, others), Unitree (G1), X-Humanoid (Tiangong Ultra), Xiaomi (CyberOne).
- Japan. SoftBank Robotics (Pepper), Kawada Robotics (NextAge), Toyota (T-HR3).
- South Korea. ROBOTIS (ROBOTIS MINI, OP3), Rainbow Robotics (Hubo).
- Europe. PAL Robotics (Spain, TALOS), Engineered Arts (UK, Ameca), Neura Robotics (Germany).
- Smaller startups and research institutions also contribute

There are over 50 companies currently developing humanoid robots with 2 arms and 2 legs. Some estimate there are over 160 if the definition includes robots with two arms but wheeled bases. If a robot is placed on an existing production line and does not need to move, then it would not need a battery system nor mobility.

Figure 21 shows the line-up of robots in 2025 from madevisual.co.

Figure 21 shows a limited lineup of humanoid robot. If you count the smaller players, [65] China leads with 60, the USA 30 and Europe 40. As mentioned, the provision of chips and opensource development software from Nvidia means barriers to entry are low, and we are likely to see hundreds of robot manufacturers eventually.

Cern Basher on X [66] has a great visual of companies and player milestones. Milestones include walking, doing useful demonstration work, working autonomously, piloting in factories and being deployed in limited use.

8.2.6 Leading The Pack

None of the robots have yet made it out into widespread deployment, but Tesla and Figure are possibly the 2 leading humanoid robotics developers. There are others in China, but there is limited visibility of who is leading the pack.

- Tesla's target is to have hundreds working in factories by the end of 2025. In the Tesla 2024 investors call, they said they expect to increase production orders of magnitude annually. That means production of hundreds of thousands by the end of 2025, a million in 2026 and tens of millions thereafter. Tesla's advantage is two-fold – it has some of the largest AI training clusters globally with an AI inference computer already deployed in over 6m vehicles. It also has some of the most efficient modern manufacturing factories. Will this be enough to maintain the lead?

- Figure has a number of robots working in their partner BMW factories and has announced building a manufacturing centre [67].
- In Amazon distribution centres, a robot called Digit from Agility Robotics is moving bins from a shelf to a conveyor. It plans to make 10,000 a year and deploy them to warehouses and storerooms all over the world.
- There have been dozens of announcements between humanoid robot developers and manufacturers. E.g. Toyota has its own in-house robot they have trained to do over 500 tasks, but has also partnered with Boston Dynamics.
- The humanoid form is likely to dominate over the next decade. We can see there will be forms that are more niche. Some robot developers will follow down those models.

8.3 Compelling Economics

Many have developed economic models even though the actual manufacturing processes and costs are still in the research and commercialisation phases. Some humanoid robots are selling for $100,000 but most expect those prices to be under $20,000 and as low as $10,000 based on the reduction in cost of sensors, batteries, actuators and computer systems.

> *Humanoid robots will enter the market at a cost-capability of under $10/hour for their labor, on a trajectory to under $1/hour before 2035 and under $0.10/hour before 2045* Dr Adam Dorr RethinkX [56]

This simple model following demonstrates the economics.

The Cost of Human Labour

- Minimum wage = $16 per hour

- Add on-costs such as supervision, holidays, recruitment, training and rehiring.
- After on-costs = $46 per hour
- Annual cost = $92,000
- Cost of five years = $0.5 million.

They work just 1 shift. The cost for 3 shifts as day i.e. for 24 hours per day and 7 days a week would be 4 times that or $2 million. That is part of the reason why 24 hour care is so expensive.

<u>**Cost of a humanoid robot**</u>

- Capital cost $20,000 robot
- Works 22 hours per day
- 7 days a week
- 5-year lifespan
- Amortised capital cost = $0.49 per hour.
- Let us add 10% for maintenance and electricity costs
- Cost = $0.55 per hour.
- Cost over 5 years = $22,000.

The FOAK (first of a kind) cost might be much higher, perhaps $100,000, but even that is less than 5% of the cost of a human worker 24/7/365.

Ark Invest and others have more complex models and show much higher returns on investment. Even these very high returns are too conservative, say others.[39] It is clear that when the capital cost falls below $20,000, the total operating cost will be less than $1.00 per hour. Where in the world are wages less than $1 per hour?

The business case is compelling. As a business, would you relocate your business to another location or country to access lower labour costs, or invest in more humanoid robots? If you are an aged care facility and struggle to get staff, would you not acquire humanoid robots to do the cleaning, bed-making, and cooking, and use existing staff to spend more time with the residents? These have major repercussions on society, and we will explore them in more detail below.

8.4 Indirect Economic Benefits

In addition to the direct economic benefits, humanoid robots signal the end of worker drudgery and management stress. Managers the world over complain that work would be easy without the hassle from workers. The indirect benefits of a humanoid robot would include:

Never late for work.
Never get sick.
Pandemic-proof.
Never quit - reducing turnover.
Work tirelessly.
Don't get injured or disabled.
Don't demand workers' compensation.
Elimination of hiring, turnover and training costs.
No Human Resources department.
Never complain about performing repetitive work
Don't need healthcare.
Can handle dangerous/hazardous tasks.
No retirement benefits.
No workplace harassment/discrimination.
No performance reviews.
No awkward/drunk holiday parties.
Don't fight with other employees
Don't go on strike.
No workplace politics.
Don't steal/sabotage.
No parking space needed.
No bathrooms, food service, cafeterias or snack vending.
Less middle management.
No labour-employer disputes.
No need to coordinate vacations.
No unexpected emergencies.
Easier to scale a workforce.
Don't need maternity/paternity leave.
Can instruct them in any language.
Bots continuously improve their skills.

Can be relocated to the next town, interstate or international without the cost of relocating family.
Always do the same quality.
Bots "happy" to work 22 hours / day.

If the job changes or new skills are needed, then can download a new app or provide new software so the whole workforce is trained to the current production status.

8.5 Implications of Form Factor

Changing from an industrial to a humanoid form factor has multiple implications. We will explore below the implications. From tasks to jobs, from sovereignty to slavery, from aged care to home care, from domestic duties to warfare, and more than twenty others listed in the following sections. Some are local, and some are national. Others are global, from economic and conflict, including warfare. And for some, interplanetary. As the book title implies, this end of drudgery is one of the biggest societal changes we have ever faced. Not just from before the Industrial Revolution; this change goes back millions of years. Back to when our sole focus was ensuring we had enough food to survive and reproduce.

8.6 Are Robots Inevitable?

We will conclude that the humanoid robot's disruption of labour is inevitable, and explore the reasons why. Rethink X released its first analysis of humanoid robots in May 2024 [27] and noted that throughout history, whenever technology has enabled a 10x or greater cost reduction relative to the incumbent system, a disruption has followed. This number alone makes the disruption of a substantial fraction of human labour inevitable. They also point out that humanoid robots today are the most expensive and least capable they will ever be. Governments are promoting the development of both state and private enterprises, and the state of robotics today stands on the cusp of full production. Since deciding to write this book two years ago, humanoid robots have advanced from actors to real products. Mass manufacture won't start until production models have been finalised, as we are still in the production prototype phase. The move from prototype products to production line requires a large capital investment, and most companies will delay ramping up production until design engineering is finalised and prototype deployment is assessed. Humanoid robots are a combination of hardware and software. Software will continue to evolve. Hardware will go through various releases. Some suggest that humanoid robots will operate similarly to mobile phones and apps. One would buy a model with limited functionality. Core operating systems would be updated frequently. Specific applications would require an additional purchase and download. Want a robot to do laundry? The laundry app does that. Gardening? Shopping? Aged care? Even with something such as a laundry app, we might see stratification of abilities. The basic app supports five major washing machine manufacturers. Others might need to pay for the add-on app.

8.7 Tasks Not Jobs

Projecting the status of robots in a decade or two is very difficult. We don't have existing humanoid robots. We are transitioning from industrial specialised robots to humanoid robots. Many jobs have not been segmented into tasks and allocated to humanoid robot tasks and human tasks.

Humanoid robots will initially perform tasks. Some say robots will take our jobs but jobs are not to be confused with tasks or especially work. [68] To clarify the task, job and work, we can look at a railway worker shovelling dirt into a wheelbarrow and pushing it into a ravine. That process is a job made up of various task, work and jobs. Work was replaced, first by a horse and cart, then by a bulldozer. The worker's job became driving the bulldozer. A bulldozer working is multiple times more effective than a human or a robot, and has been enabled by fossil fuels and, more recently, renewable energy. To be clear, robotics is not about work. The multiple tasks of driving a bulldozer have been slowly replaced, even before robots. These days, the task of levelling the road has been augmented by using lasers and control systems to make sure that task of levelling, is done by a machine. Other tasks will be taken on by other robotic systems and in time all those tasks are completed by dedicated autonomous systems. The installed base of bulldozers will preclude an immediate swap-out, and it could be cheaper to purchase a humanoid robot who can sit in a bulldozer and drive it. Alternatively, there are many instances where a job such as driving a mining haul truck is now carried out by an operator sitting 2,000 km away using remote control services. It is entirely probable that a humanoid robot will use existing equipment. The washing machine is likely to continue without much change. The washing machine design might be changed from being operated by humans to being operated by humanoid robots, but the installed base of washing machines precludes rapid change.

The automotive manufacturing industry is one of the most automated globally, yet there are still 10 people for every robot. At the latest Tesla Shanghai Gigafactory, they replaced the manufacture of the rear and front with castings produced by the "Gigapress." These machines reduced the number of robots from over 1000 to 500. The factory can produce over 1 million cars per year, yet there are still over 20,000 employees. That's 20 people per industrial robot. What do these employees do? Does it suggest that multiple tasks are not economically feasible to automate? Are they engineering jobs or are they more mundane manual jobs? Tesla has recently implemented a robotics function at its Texas factory. Previously, cars came off the assembly line, and then over 30 workers drove them 2km to a final checking station. That task is now gone as they programmed the autonomous driving to the final inspection building. It won't be long, and those autonomous cars will drive themselves to their customers or to the nearest railhead to be transported across the country.

In contrast, a health care worker, a home worker or a worker in a factory assembling a drill all have multiple tasks in their job. They could offload individual tasks to a humanoid robot. The task of picking up a drill, the components, and assembling them may be one task. Testing that drill and stacking it into a box would be another task. These tasks will be given to the robot as they are discrete and repetitive. The worker will have other tasks as part of their job, such as packing up a pallet with drills and shipping to a distributor. Or that worker may have multiple jobs within that business and is deployed one week making drills and another week making a chainsaw. These tasks can be learned by watching someone else perform them, by watching a video, by downloading an app from the "Robo App Store", or by having sufficient AI cognitive skills to complete the task without training.

If we consider a professional dishwasher job, only part of the job involves actually washing the dishes. The plates are collected from the table. Food scraps were removed, and the plate was placed in the dishwasher. The dishwasher robot (dumb) washes the plates. Then the tasks of emptying the dishwasher and putting the plates away in a storage cupboard are performed by the humanoid robot. The human dishwasher may also oversee food deliveries and storage. Even simple jobs in a sector with thousands of professional dishwasher positions involve multiple tasks. I would predict that the takeover of tasks will follow that path of gradual, then sudden.

The Traceys of this world are in for a shakeup. How many of the one billion well-off people decide that they could have help with washing, cooking, cleaning, tending a garden, and providing supervision and after-school care? Would they? Remember that they do buy cars for $20,000 to $80,000 and use less than 4% of the time? Could they provide childcare at a fraction of the cost? Or provide robots for the menial tasks that elderly or disabled people still require.

8.8 Jobs Will Be Gone

One of the misunderstandings about AI-controlled robots, both industrial and small humanoid, is that they enhance productivity and that there will be MORE jobs. As we saw in the previous chapters, with 4 out of 5 jobs in the service sector, AI could be a "productivity tool" and increase worker productivity. We will explore how tasks and jobs in the blue-collar/physical labour sector differ. Some still think we will see job creation from AI and robotics, and that new jobs will be created as quickly as or faster than job destruction. It can be discouraging to see otherwise bright and mostly sensible people parroting the lines of "no job losses." If there is one major point about humanoid robotics in the blue-collar/manual sector, it is that humanoid robots will replace workers. There will be humans to supervise ten or 20 robots but it is a far cry from more jobs. That may be temporary, but the very essence of robots in a humanoid form is that humans can step back into that task, or as mentioned they could be called cobots or co-existing robots.

AI and robots will be more capable than humans in EVERY way within a decade. The marginal cost of virtually all labour is headed to zero. This is not the Internet or the automobile all over again. There won't be 5 billion "influencer" jobs in 2040 once 25 billion robots are running GPT9, even if it isn't fully sapient AGI in every one of them.

Will humanoid robots take our jobs with the automation of labour? Will it mean higher unemployment? There are two differing guesses, and you make up your mind based on the next few sections.

A white paper from ARK Invest [69] says no for two reasons:

- Automation can increase wages.
- Automation can transform non-market activity into commercial activity.

Ark Invest was focused on the knowledge worker. Humanoid robots are in the blue-collar sector. It is unclear yet how this disruption will unfold. Labour is a key component of our humanity. Where will robotics make the most inroads?

The fear of mass layoffs is currently raised in the mining sector as renewable energy decreases demand for coal and gas. Those employed in Australia's coal and gas industry number fewer than 38,000, far fewer than the 100,000 who work for McDonald's or the 390,000 nurses in Australia. Humanoid robots may not displace those nurses in the first decade, but it's pretty certain that the number of fast food employees will diminish. By how much? It's speculation, but 70% or more seems a minimum. Next time you visit a fast food restaurant, observe the processes and evaluate how many of those repetitive tasks could be performed by a humanoid robot. Putting patties onto the grill. Setting a time. Removing them after 2 minutes 40 seconds and placing them on a bun? Calling out numbers? At the very least, a robot could clean the bathrooms and wipe tables, though one might want two of them, or at least include a cleaning step between tasks.

The view that human robots won't reduce jobs is deeply, deeply discouraging short-sightedness, according to Dr Adam Dorr of RethinkX [56]. He says jobs will be gone. It is just a matter of time. It will start slowly with tasks, but eventually, all the tasks in a large number of jobs will be gone. Will others come to similar conclusions? It is not necessarily bad, but it is inevitable as technology and economics drive disruption, even over politics and belief systems. We will discuss in more detail in a later chapter.

8.9 Everywhere

People's views on humanoid robots need adjustment because they are inconsistent across sectors. While editing this book in Bali, I had a discussion with a very intelligent European over breakfast. He has just started a small restaurant as part of his foreign visa permit for Indonesia. Under that visa, they can invest and start a business, but not work in it. Not even picking up a spatula to show the chef how to flip a burger! Here is a society where even training and taking a job from a local is prohibited.

On the one hand, he had been explaining how he was using ChatGPT, but then he said that autonomous driving would never happen in Bali because of the chaotic nature of the roads. To which I replied, “People drive these roads, and most use motorbikes.” Why would you think AI and robotics could not drive better than humans with zero accidents? Silence. We moved to a discussion of cooks and food preparation. He said people wanted to get their food in his cafe from a real person. So, I challenged him again. If humanoid robots can do the tasks involved in cooking a burger 100% reliably, provide consistency of food with consistent quality, why would you not change if the economics stacked up? If he could train those two cooks once, without lifting a spatula, what was stopping him? The wage costs in Indonesia are 3.5 million rupiah per employee per month ($260). The wages bill for an employee over five years would be $18,000 plus oncosts. Why would he still employ those workers if a humanoid robot cost the same? More reliable. No employment hassles. No oncosts. Work 16 hours a day providing food service. No time off to go to family and cultural ceremonies. His response was that a robot would remove the personal interaction with customers. To which I asked if he provided a food service or a personal counselling service. It makes economic sense to employ one person to talk to customers and provide interaction that most humans appear to need. They could even serve the food to improve interaction. The robot: food preparation and cooking. In Japan and other countries facing labour shortages, robots deliver food. Clean the table. That scenario is probably the wrong way around. Should humans do the interaction? Restaurants operate with a minimum staff. Gone will be the days when restaurants had about as many wait staff as diners.

The disruption to labour will create an entirely new labour system, not just in Western countries but also in the less developed nations.

8.10 Market Size

Any conjecture of the market size for humanoid robots is a wild guess at this stage in the technology development curve. When one builds models that examine the value proposition, as we have done previously, the numbers are so large that they look ridiculous. Are they ridiculous? Probably not, especially if we base our estimates on the rise in industrial robot adoption. The humanoid robot market is more like the mobile phone market. In 1985, McKinsey Consulting provided AT&T a forecast for the number of mobile phones, and the 15-year prediction was that the total number of phones would be 900,000. The actual number? 108 million. Wrong by 100 times. In 2025, over 1 billion new phones are expected to be sold, resulting in more mobile phones than people. The price elasticity is the critical data point about guessing how many humanoid robots we expect by 2045.

- If the robot price were $50,000, one could expect a similar number to cars, which is about 1.5 billion.
- If the price is more like a laptop computer, the numbers could be as many robots as people (8 billion).
- The cost per hour will be 1/10th of the cost of human labour. It is anyone's guess.

8.11Mass Unemployment

Countering the fear of mass unemployment is data that shows most countries are experiencing population decline and an aging population. Australia's centenarians (those 100+) have grown from 1,300 in 2000 to a projected 50,000 by 2050. Population growth is discussed in relation to Earth's carrying capacity in a later chapter. Birth rates are below the replacement rate. The Earth's population is increasing because we have better health for the elderly, particularly in Africa. Worker segments in most Western countries are increasing through migration, but this has political ramifications, especially in high-migration countries such as the USA and Europe. Because humanoid robotics is new, we have no track record of what will happen, and it is helpful to look at other disruptive technologies.

When computers were introduced, many said we could save paper. That was a joke. The amount of paper has increased by 70% since 1990. For the first two decades, everyone had access to unlimited fonts and revisions, and storage of files and documents became big business. Only in the last few years has paper use slowed down, even though the printing industry is still growing at 5% annually. We need to look closely at which sectors are likely to lose jobs first. Morgan Stanley [64] maps out 21 industries and forecasts which may lose up to 35% of their workers by 2050. Their data is useful as a framework but little more than conjecture, given the uncertainty of technology and a long time in the future. There are some obvious jobs at risk. Six years ago WEF [70] analysed the share of time spent in tasks that could be automated. Not much has changed in the following six years with regards to:

- Workers moving boxes or materials around a factory.
- Taxi and rideshare drivers.
- Hospitality sector jobs. Workers spend much of their time doing physical tasks in a predictable environment. Bar workers, waiters, cooks and cleaning staff still do similar tasks.
- Retail sector. We have seen self-service systems appear, but shelf stacking and restocking are jobs that humans still do but would appear to jobs robotics could do.
- Inspection. Inspection and quality improvement will be among the first to go. Humans get bored and distracted.
- Agriculture. We are already seeing automation taking out the drudgery – autonomous tractor driving and crop management.
- Transportation and warehousing. Industrial robots have made inroads but there are still many tasks that require humans.
- Mining. Many jobs are repetitive.

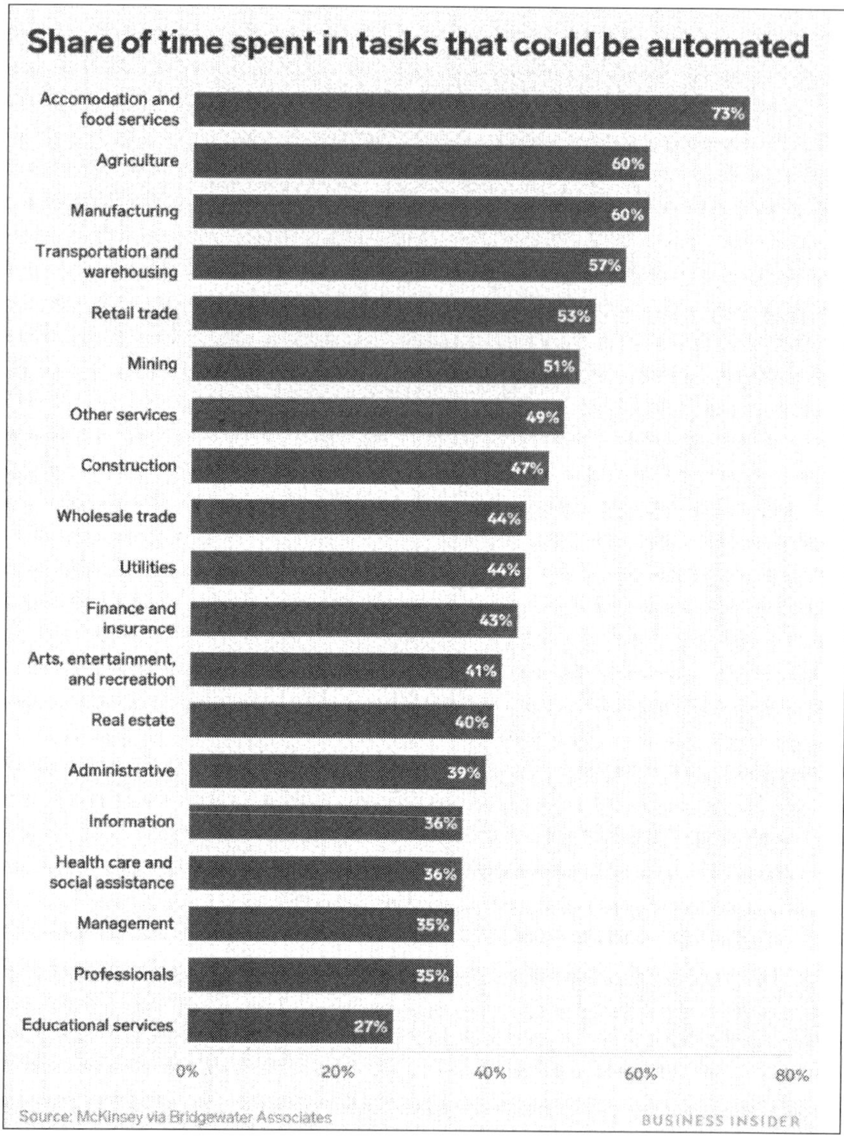

Figure 22 Industries potentially automated

Throughout history, automation has sparked fears of unemployment, even though technology appears to have been a net job creator over time. Yet, with each wave of innovation, many say, "This time is different". Ark Invest argues that it will be no different this time, that automation increases wages and salaries, and people simply move from one industry to another. RethinkX disagree. They think this is a different disruption.

8.12 GDP Increases

GDP measures the monetary value of final goods and services. I.e. those that are bought by the final user and produced in a country in a given period of time, e.g. a year. It counts all output generated within a country's borders. Currently, the GDP is based on human jobs, so what happens to the measurements of GDP when humans no longer do jobs? The robot will require electricity and maintenance, but not food, waste, or water. Given that it can work 22 hours per day, task productivity is 20 to 30 times higher, with no marginal cost. With the same number of people, will GDP increase by 20 to 30 times?

Neither do robots buy goods and services. How many robots? If the number of robots exceeds 1 robot per person, then GDP will rise rapidly, as the output of goods and services will no longer be constrained by the number of people. Note that GDP is a measure, and we will probably need to come up with a new measure that accounts for the reduction in the cost of energy and labour.

With tasks taken over by robots, what will humans do when robots not only do the tasks but also start doing whole jobs? We will look at this in a following section.

8.13 Trade Disruption

Trade disruptions have prompted many global companies to shift their production and sourcing locations. Boston Consulting reports [71] more than 90% of the North American manufacturers surveyed relocated some production from China in the past five years with a similar percentage plan to make such moves in the next five years. And this is before the USA embarked on trade tariffs in the 47th presidency. Mexico, Southeast Asia, India, Turkey, and Morocco are among the most competitive destinations. Apple and its ecosystem have directly created 150,000 jobs in India and are planning to move half of its supply chain to India from China along with over 0.5 million jobs. [72] Supply chains are important, but even more so are geopolitics, high tariffs, labour costs, skilled labour and energy costs.

Humanoid robots and renewable energy will radically disrupt these plans. It makes sense if the manufacturing moves closer to the market, or if trade agreements distort markets, or if there are large domestic markets in those countries. The reduction in cost of labour, anywhere in the world, at less than $2 per hour at an equivalent quality, is going to disrupt where people get work done. E.g. an Australian fish farm grows salmon in Tasmania with colder, cleaner water. After harvesting the salmon, it is sent to a factory in Thailand, where a low-cost labour force fillets the fish, and the fish is then packed and quick-frozen. The frozen portions now return to Australia to be sold as Australian-produced. With a labour force of a few dozen robots, working 22 hours per day, with absolute quality, that fish is processed in Tasmania. The company now has control of the supply lines with zero risk of shipping disruption. The energy from their renewable energy is as cheap in Tasmania as in Thailand. Furthermore, if the quantity increases, the farm buys or purchases more humanoid robots. A quick download of the fish filleting app and the new robots are ready to go to work. Zero training, 100% quality, and 100% control of the whole process.

Manufacturing could become sustainable within Australia if labour costs tend towards expected levels with the introduction of humanoid robots. Robotics level up the labour market. By 2035 adding one million people to a nation's workforce might cost $100 billion and take 20 years, whereas adding one million humanoid robots to its workforce might cost just $10 billion and take a single year.[56]

Manufacturing location now becomes localised. A humanoid robot is as cost-effective to operate in Wagga Wagga, Sydney, or Shenzhen. Supervision might be cheaper in Shenzhen, but other costs will be much reduced with local manufacturing. Trade risks, currency exchange risks, shorter lead times, and maintaining quality in-house all contribute to lower sovereign risk. Manufacturing can be moved back onshore. Politicians have not yet woken up to this sovereignty upheaval. The current political approach in the USA for high tariffs to generate increased employment is counter to where technology is taking us. Any new factory will be based on tasks performed by humanoid robots, as quickly as manufacturers can build them. What happens to those promised jobs? Will not be there.

8.14 Home Care and Living Aid

Australia introduced the National Disability Insurance Scheme (NDIS) in 2013 to provide services and support to people with disabilities. The cost of the scheme has risen from 1.7% of GDP and is projected to exceed 2.1%. The government is rushing to control expenses while extending services to over 600,000 individuals. In comparison, total military spending is around 1.9%, a level that has remained stable since the end of the Cold War. Governments claim they want to cut costs, but no government is willing to risk losing that many voters. The $48 billion price tag translates to $80,000 for each recipient. A significant portion of the cost is recurrent labour for services including daily personal activities, transport, and help with household tasks such as cleaning and home care. A cost of $2 per hour appears appealing when contract cleaners charge $80.

One of my grandchildren has Down syndrome and requires additional care, currently provided by her parents. But as her parents age, who will provide the daily care? Cooking. Clothes. Transport. Shopping. Exercise. Education. Provision of care to those 6 million globally with Down syndrome takes an enormous stress off their carers and would enable a robust 24/7 care system for all.

8.15 Aged Care

In the chapter on population, we explore some of the parameters of the aging global population. We also examine the challenges of keeping those healthy and well, rather than a crippling burden from dementia and other elderly illnesses, in the health chapter. We had 1,300 centenarians in Australia in 2000, over 6,000 in 2018, and projected to be over 50,000 by 2050. Who will provide appropriate care? Migrants? Australians? Humanoid robots will be able to do these tasks by 2030, and most certainly by 2045. The tasks that can be done by humanoid robots include:

- Assisting elderly and disabled individuals with daily tasks. Cooking. Cleaning, personal care and individual attention.
- Individual meals. Food in institutions is notorious poor. Boarding school/hospital food has a well-earned reputation. Humanoid robots

should be able to prepare individualised food and ban bulk cooked food forever. It is not that we need to have 5-star Michelin food quality every day, but that is the indirect impact of highest quality, every time and on time, always.

- Humanoid robots are increasingly being developed and used to interact with the elderly, offering potential benefits like companionship, assistance with daily tasks, and emotional support. Robots engage in verbal and non-verbal communication, participate in activities, and even provide reminders and prompts. There are risks, especially with patients with dementia. [73]

8.16 Education

The role of AI and robots in education is a hotly debated topic. Humanoid robots could do some of the jobs that people in these sectors do. Which tasks can be or should be offloaded is unclear, as the emerging technology is not yet mature enough. In a decade, anything is up for debate.

- Child Carers: Provide care and supervision for children in various settings. High demand, but often low pay and challenging conditions, contribute to career shortages. There are two different reasons for child carers. Firstly, childcare allows parents to get back to work, and secondly, there is good evidence that children can benefit from early education and socialisation.
- Early Childhood Teachers: Educate and care for young children. The shortage today is due to increasing childcare needs and demanding work conditions.
- School and University teachers are struggling to bring AI into the classroom and to find where AI fits in the new society. A recent report says Caltech has a program in which AI is integrated into every part of education. Will this be the new norm? [74]

There is a view that school is a place to babysit children, so they can learn to be bored. As they leave school, they often take a job where they continue to feel bored. In summary, our education system is designed to prepare a portion of future workers for monotonous and repetitive tasks. With those dirty, dangerous, and drudgery jobs gone, what is the role of education? That is too broad a subject here, but questions need to be asked.

8.17 Population and Migration

Australia has a major shortage of skilled workers and hence moderately high migration levels. Other countries have even higher migration to meet skilled job vacancies. Humanoid robots will free up existing labour and may reduce the need for skilled migrants. Some skilled areas are not yet suited to humanoid robots, such as construction managers, civil engineering professionals, and IT systems analysts, but there are other skilled professions that humanoid robots could relieve pressure on by reducing some of the tasks in those jobs.

- Registered Nurses. Many of the tasks are mundane and repetitive. There is a need to provide healthcare in various settings, and demand will increase due to the aging population and burnout in the profession.
- Electricians. Most of the job of installing and maintaining electrical systems could be done by humanoid robots. This is an essential trade with a shortage of qualified workers, especially in specialised areas, and many of the tasks could be offloaded.
- Chefs prepare and cook food in restaurants and other settings. The hospitality industry struggles to find and retain chefs, particularly those with specific cuisine expertise. Humanoid robots could prepare the most exotic and labour-intensive dishes consistently, every time. Some of the chef work is artistic, but as chefs are trained by watching others, it seems entirely possible to move many of the tasks to humanoid robots.
- McDonald's have deployed robots and AI for food ordering and delivery, but has had some comical mishaps. They appear to have withdrawn to regroup. Expect that with the rapid improvement in AI and humanoid robots, they will have another go. The economics are too compelling; they can improve their service measures, and they already have well-defined processes.
- Metal Fitters and Machinists: Produce, assemble, and repair metal parts and machinery. The manufacturing and engineering sectors face a shortage of skilled tradespeople. Many of the tasks could be learnt. The mechanic's role becomes more of supervision than doing.

8.18 Autocatalysis

In chemistry, autocatalysis is a process in which the product of a chemical reaction catalyses the reaction itself, thereby speeding up its own formation. This "self-catalysing" effect can create a positive feedback loop, leading to an exponential increase in the product over time. This term applies to humanoid robots because they will be used to manufacture other humanoid robots. Examples exist in the past. The coal mining industry really started growing when coal was used to mine coal. Prior to that step, small "pit" ponies were used to haul coal out of mineshafts. When steam engines were used from the mid-1700s, coal production exploded. The more robots are used to manufacture other robots, the faster the industry will scale. Between industrial robots and humanoid robots, factories will be able to churn out human robots at an ever-increasing rate. Scaling up production rapidly is challenging for most businesses. The complexity for supply chains, workforce increase and training means even a 50% increase in production is very difficult for large producers. Tesla has said it aims to produce a million units in 2030. Most of these will be made by robots.

8.19 Cheaper Products

Goods and services will become cheaper. We don't see the cost of goods and services decreasing today. Cost of living is a major political and social challenge in most countries, from post Covid to today. Will the next ten years reverse that trend? We see how AI is changing the face of services, with chatbots and AI agents already exerting downward pressure on service costs. With goods, there are only 4 components of price.

1. Cost of the raw material, down to the mass of the raw material. If there is 1kg of steel in an item, the cost of the finished product will tend downward to the cost of that steel. It can't go below that. Reducing the weight of the product from 1 kg to 0.5 kg will reduce the cost proportionally.
2. Cost of the energy to transform those elements into consumer goods. If the Australian ARENA goal of 30/30/30 is met (30% solar panel efficiency and $AU0.30 per Watt installed by 2030 [1]), then the wholesale cost of electricity will be under $US15/MWh. This is

approximately one-fifth the current average global electricity cost of around \$80/MWh. Moreover, Stellar energy, as suggested by RethinkX [28], will be an era of superabundant electricity where electricity will tend to zero cost.

3. Cost of labour. Substitution of labour by capital is a common method for reducing costs, in which automation and industrial robots are used. Lowering the cost of labour from \$20 per hour to under \$2 per hour drives costs out of the product.
4. Profit margin. General economic theory suggests that competition will drive down profit margins, although we will see in the population chapter that excessive inequality with profits accruing to fewer and wealthier individuals.

The reduction in cost of goods will not happen for some time. Even with energy costs and labour costs falling, do not expect prices to fall in the first five years. The complexity of commerce and the drive by major corporations to increase profit margins mean it will take time for prices to fall. It may simply be that prices do not rise as fast as they did, removing pressure points for ongoing price increases.

8.20 Quality of Goods

Goods are often differentiated by quality. A handbag, a set of drills, or an item of clothing is often categorised as luxury, well-made, or of cheap quality. You can buy a shirt for \$200 or buy one for \$6. Where does that price difference arise? Is it the fabric quality or the sewing quality? With humanoid robots, it will be technically feasible that once you have a task defined, e.g. sewing a seam, the robot will sew that seam exactly the same for every garment. Perfectly. There will be no seconds, no quality range. Everything will be 100% perfect. So how do the goods get differentiated? There won't be \$2 discount stores, as everything will be the best-produced goods.

Some of the price variation today is due to labour and energy costs in the factory, but a majority is how companies take profit during the supply chain. If labour costs are essentially free and energy costs are free, how does the item get differentiated? Standardisation of quality is an unexpected implication when humanoid robots begin to be part of the manufacturing cycle, and changes in pricing will be an interesting outcome.

8.21 Status Symbol

There will be a change in the status symbol that humans aspire to. Ada in Africa aspires to a bicycle that relieves the walking she does. Tracey aspires to a better car or bigger house.

The rise of personal public transport via robotaxis and ride-hailing services will also devalue the very concept of car ownership as a status symbol. Vehicle manufacturers used status as a key purchasing decision during the 1950s with the marketing slogan of "keeping up with the Joneses"? What status symbol will replace the status of vehicle ownership? If car ownership is equated with wasteful expenditure, then using personal public transport becomes the sensible decision, and owning a car that sits out the front of the house and depreciates in value becomes stupidity. Car manufacturers won't like that and will continue to promote status, but how long will that last until they go broke selling fewer and fewer cars?

Households may well look to humanoid robots as a status symbol. Who in the Western world would not want a home cleaner, a clothes manager, a cook and dishwasher and a home gardener tending your own vegetable garden? With the right sort of app, they could even provide everything for a dinner party. Dinner parties have gone out of vogue – too much work. Dinner parties and other social events become more affordable when you can rent a humanoid robot at $20 per hour. Hire a few extra humanoid robots to be the waiters, and a couple of chefs to cook fine food instead of restaurant-prepared food delivered via DoorDash/Uber. In the USA, over 70% of food is pre-prepared and 30% of households do not cook. We can envision a business opportunity to utilise humanoid robots for these events. Novelty will turn into acquisition.

8.22 Dirty and Dangerous Jobs

Dirty, dangerous, and drudgery jobs will be the first to go because of costs, followed progressively by increasingly complex tasks. The most likely first application is one where there is a high return on investment or labour shortages. Any job not done by humans because it is too expensive, too dangerous or too tedious becomes feasible with humanoid robots. Cleaning streets, sorting garbage, picking food crops? The flexibility of a robot in a glasshouse food production means it can pick strawberries, cucumbers, pack them, sort them and obviate the need for dedicated robotics. At a cost of 10% of a human. Some jobs are not done due to their dangerous nature. For example, cleaning out industrial tanks is done by specialised workers. Toxic chemicals or gases create hazardous working conditions. A humanoid robot can go into those environments without risk.

Australia's current work-related injury rate of 3.5% is the proportion of people who experienced a work-related injury or illness in the previous 12 months. Australia's rate is about one-third of the global rate of 12.1%. [75] It is not just the 1,880 traumatic injury fatalities over the past 10 years – it is the 1,165,100 workers who have made a serious workers' compensation claim involving more than one week of working time lost. Fatalities are serious, but it's the loss of productivity that is crippling the country. The cost to owners with increases in insurance coverage is huge. Over 76% of work-related traumatic injury fatalities and 61% of serious workers' compensation claims occurred in just these following 6 industries:

- agriculture, forestry and fishing
- public administration and safety
- transport, postal and warehousing
- manufacturing
- health care and social assistance
- construction.

Vehicle incidents account for the most significant proportion of worker fatalities with 42% (84 fatalities) but falls from a height at 15% (29 fatalities) are also a major problem.

An example of a future is a small gutter-cleaning business. The owner employs 3 staff members and is part of a network of over 80 franchisees in Australia. These workers get up onto roofs with hoses to suck out the leaves and debris in gutters, reducing storm damage to buildings. Drones are not yet sufficiently powerful and would need operators to ensure the gutters are clear. It appears the primary staff qualification is being willing to turn up for work! There is training, but the biggest challenge is to find people willing to work at heights.

Workplace Health and Safety insurance premiums are astronomical for this business. A $50,000 humanoid robot would be cheaper than a worker at minimum wage for a year. If a robot falls off the roof, a quick repair and they are back in business. Every job would be done to an expert level. With minor adaptation, the humanoid robot could also work on wet roofs, currently illegal under workplace guidelines, and lead to a 30% increase in output per shift. Like elsewhere, robots change the framing of the problem. They could work 22 hours a day or overnight for buildings such as schools. Weekends become part of the past. Practices like these are currently uneconomic. Robots are a fixed price and the longer they work, the better the return.

In factories, safety barriers are put up around industrial robots to prevent human error. Humans tune out and get fingers amputated or crushed, while there are other people who disable safety switches. Underground mining is particularly high-risk, with risks of falls, gas, and equipment failure. Consider the return on investment in underground mining. $2 per hour versus $150,000 per year per shift for workers. Those mine operators will be lining up to replace most of their workforce. That does not even start to address the liability for black lung disease and other costs incurred.

8.23 End of Slavery

Labour and slavery have been with humanity for millennia. Evidence of slavery predates written records; the practice has existed in many cultures and can be traced back approximately 11,000 years. It appears to have arisen to the conditions created by the invention of agriculture during the Neolithic Revolution. Economic surpluses and high population densities made mass slavery viable. [76] The conquering country kept its enemies captured in war as slaves.

Slavery operated in the first civilisations, such as Sumer in Mesopotamia, dating to 3500 BC. Slavery featured in the Mesopotamian Code of Hammurabi (c.1860 BCE), which refers to it as an established institution. Slavery was widespread not just in the ancient world but in more recent times. We are all aware of the slave trade in the settlement of America. Over 13 million people were captured and sold as slaves between the 15th and 19th centuries.

The Industrial Revolution accelerated the trend toward drudgery, characterised by low wages and long work hours. We have different names for this slavery, but forced labour, debt bond, forced marriage, and human trafficking are all forms.[77] Some Australians like to think Australia never had slavery, but the history of convicts and treatment of Aboriginals is archetypal slavery, as much as it was in most known civilisations.

Ada belongs to a culture with widespread slavery—desperate individuals fleeing poverty and hunger will go to extreme lengths. Tracey believes there is no slavery in her local town, but rest assured that the rise in immigration has led to businesses exploiting workers through wage theft. Much of the Western economy is supported by low-paid and sometimes exploited workers. Economic and climate refugees are particularly vulnerable. Slavery should have been relegated to history, yet it remains a modern-day phenomenon, says the human rights group, Walk Free, which estimates that 49 million people are in modern slavery. They define slavery as being forced to work against one's will; being owned or controlled by an exploiter or "employer"; having limited freedom of movement; or being dehumanised, treated as a commodity, or bought and sold as property.

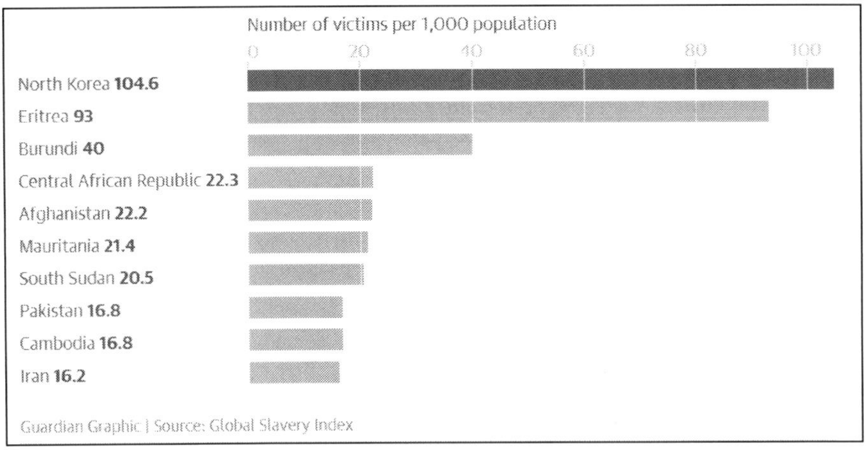

Figure 23 Slavery victims in top 11 countries globally.

Figure 23 shows the top two countries are North Korea and Eritrea, with about 10% of their populations in slavery. Still, many of the next nine countries have rates exceeding 2% of their populations, according to the UN. Meanwhile, India, China and Pakistan are home to the largest number of victims in terms of pure magnitude. Women and girls are the most vulnerable, with many being exploited by practices including forced labour, forced marriage, debt-bondage and domestic servitude. One in three slaves is a child. Fifty-four per cent are female, with approximately 27 million in forced labour and 22 million in forced marriages. The AntiSlavery group highlights that workers produce $350 billion worth of goods we use. Platforms can help companies report on slavery in their supply chains. Other estimates suggest that slavery affects about one in 200 workers.

When humanoid robots are cheaper than slavery in those countries, what will be the primary reason for slavery?

8.24 National Plans

Given the implications of lower costs and higher GDP, investing in humanoid robots should be seen as in the national interest. One country has taken this challenge seriously. China's Ministry of Industry and Information Technology, along with 17 other agencies, created a new action plan called the "Robot + Application Action Plan." This plan outlines 10 industries in which it will develop robotic systems for, with an overarching goal for the China's robotics industry by 2025. [78] China hopes to accelerate the adoption of robotics in manufacturing, agriculture, architecture, logistics, energy, healthcare, education, elderly care, commercial community services, and emergency and extreme-environment applications. If China achieves the same level of dominance it has in manufacturing, EVs, batteries, and solar panels through its planning and investment, expect to see rapid progress. Other countries have their eyes on AI, but not on AI for humanoid robots.

8.25 Military Use of Robots and AI

We are currently in what some describe as an AI arms race. Along with robotics, we need to include AI, not just AI itself, but also in conjunction with AI-controlled humanoid robotics, AI-controlled drones, and AI-controlled "conventional" weapons, including missiles, artillery and vehicles, including tanks and ships. Will we reach a stage where AI systems or humanoid robots take on the fighting? There are enormous implications for global security and the nature of future AI-driven conflicts across all aspects. We are seeing an arms race for AI. Billions are invested. Spotify's CEO was an investor in Helsing, a $700 million European defence technology startup that makes software using AI to analyse large amounts of sensor and weapons-system data from the battlefield to inform military decisions in real time. It is also developing its own military drone. Venture funding in Europe's defence, security and resilience sectors reached an all-time high of $5.2 billion in 2024 and is driven by the war in Ukraine. The areas of investment include:

- Decision support systems.
- Logistics and maintenance.
- AI disinformation campaigns.
- Target resolution.
- Autonomous weapons systems.
- Cyber warfare.
- Intelligence gathering.
- Tracking and monitoring of civilians.
- Control of systems such as large fleets of AI-controlled drones.

Humanoid robots present a dangerous future. Most humanoid robots are designed to be no stronger than humans, yet it would take little to modify those robots to have superhuman strength. As we have seen in the invasion of Ukraine by Russia, $5,000 drones can destroy multi-million-dollar aircraft and ships. Humanoid robots can be manufactured in almost unlimited quantities, whereas the recruitment or conscription of a nation's young people is constrained. We need to recognise that deploying humanoid robots by rogue and not-so-rogue nations is highly probable.

The Ottawa Convention of 1997 on land mines is relevant to understanding countries' attitudes towards banned munitions. Treaties signed in peace are often vulnerable during times of war. Ukraine banned landmines, but Russia did not, and we now see the use by both parties as the war intensified, [2] with survival being everything. A treaty on the use of lethal autonomous weapons is unlikely to be signed. Should Australia or other countries invest in humanoid robots for the military? OpenAI has signed a $100 million contract with the U.S. military, and moving from software to a full system will be a small step forward. Australia spends as much on the military as on road construction and maintenance. Using robotics in the military would free up resources for other uses. Or make the 2% of GDP go a lot further when you build up a fleet of humanoid robots. Humanoid robots won't require hours of parade ground discipline or fitness programs. And despots such as Trump, Putin, Xi and others won't flinch to replace soldiers with drones.

8.26 Rapid Adoption

The speed of AI change in the services industry is directly related to improvements in AI. The only constraint is how rapidly it can be deployed into businesses. Most software can be within days or months, and the limiting factor is generally the speed at which staff can learn new systems. In contrast, AI for robotics and machines will be determined by the manufacturing capacity of the robotics companies. Once a viable humanoid robot is working, then adding new tasks is cumulative.

We fully expect a gradual increase for humanoid robots. We will continue to see them promoted, but won't see them in general use for 3 to 5 years. Robots will initially be used by Tesla and others for their own internal use, in factories. Tesla employs 140,000 people and even in the auto industry, the most robotic industry, many tasks still require the flexibility of a human. Tesla says that by the end of 2025, they envisage thousands of robots working in their factories, although that was probably Musk's overoptimism. Most innovation will spend more time in prototyping and testing. It is cheap to refine at the prototype phase and really expensive at the production phase. If something goes wrong with a production process, the cost of recall is three times the cost of the initial product. It is probable that robots will make new robots. (Autocatalysis). By the end of 2027, Tesla expects to scale to EV production levels, i.e., a couple of million vehicles per year. By the end of 2030, tens of millions. Other manufacturers in partnership with CATL, BYD, BMW and Toyota will be looking for similar levels of deployment as the business case is so compelling. Robotic companies have or will partner with manufacturing giants - as Boston Dynamics has with Toyota. New trucks will be autonomous. We may see humanoid robots driving trucks. We will definitely see humanoid robots making trucks. We may even see humanoid robots driving older trucks as the labour cost of trucking is the 40% of the cost of trucking, more than fuel (26%). As businesses review costs, they will choose those with the highest returns. Today, automation is extremely capital-intensive because the environment must be modified to accommodate robots. Humanoid robots will work in a human environment, with no change. They may be less specialised, but as generalised robots, they can do multiple tasks.

Some robotic companies are already focused on the consumer market. If anyone has tried to hire a house cleaner, they can testify that labour is difficult to source.[79] Goldman Sachs forecasts that 2% of global elderly care could be provided by humanoid robots by 2035. That is less than 10 years away.

With the cost of labour from humanoid robots a tenth of that of real people, businesses everywhere will adopt humanoid robots. Apart from the cost of humanoid robots, little other investment is needed. Returns will be immediate. Adoption will rise exponentially, to tens of millions by 2030, and billions by 2040 [80].

With open-source robotics software, dedicated computers for humanoid robots, and investment by China and other countries to accelerate innovation, it is likely that there will be humanoid robots with limited functionality for both businesses and consumers within 3 years. Humanoid robots will become the status symbol of the 2030's and drive an insatiable demand.

8.27 Conclusion of Robotics

In previous chapters, we looked at AI development over the past few years and its future. We will also look at transportation in the next chapter. AI intelligence is increasing at multiples annually and shows no signs of slowing. We have also examined changes in the white-collar knowledge worker sector and the likelihood of a 140% increase in productivity. In this chapter, we face a different question. How fast will humanoid robots, driven by AI, displace labour? All blue-collar jobs are risky. It will be widespread but may take 2 decades.

My prediction for robots draws inspiration from the novel The Sun Also Rises.

> *"How did the character go bankrupt?" "Two ways. Gradually and then suddenly"* Ernest Hemingway, 1926

The scope is unlimited, as will be the labour. The clear conclusion is that there is little technical or economic things that will slow down replacing all physical labour with humanoid robots. By now, we should appreciate that AI-powered humanoid robots will be one of the most significant changes since mankind transitioned from hunting and gathering to tilling the land. Humanoid robotics will undoubtedly spell the end of labour work and drudgery as we know it. Impacts will be much sooner than most think – within 5 to 10 years and more profound by 2045 – just 2 decades away. We are already recipients of products made by robotics, but they are primarily used in factories. We are witnessing the emergence of unsophisticated robots in various areas, including hospitals and restaurants in Japan. My guess is that a Chinese manufacturer will design a home humanoid robot that becomes a cult. Everyone will want one. With a vast training network, abilities will improve rapidly, and an App/Robo store will offer tasks for just about anything. For real work, it may take a little longer. Businesses will buy or lease all available humanoid robots, as the economics are compelling. No person will be able to compete in the "blue-collar" labour market with a minimum wage of more than $16 against a humanoid robot at less than $2 per hour.

Humanoid robots are here now, and they will never be more expensive or less capable. As each day passes, their rate of improvement puts Moore's Law to shame. Their physical improvement will be slow and steady, but their intelligence will advance at the same rate as AI's. We are at an earlier stage of disruption than with energy or autonomous electric vehicles. Industrial robots will continue to be deployed, but humanoid robotics has barely moved beyond the prototype phase. Their ultimate potential is hard to imagine.

There are positives. The low cost will mean that labour can be deployed in areas that are now uneconomic. The availability of near-zero-cost labour would allow us to reforest vast areas of land that are currently not viable for reforestation. Clean up waterways. Build roads and rail systems. Build more housing to house the homeless. Utilise humanoid robots to manufacture and install new solar, battery, cooking, and water systems, aiming to bring 800 million people out of energy poverty.

Unfortunately, we also see the potential for military use. What we must avoid is countries that use the technology to wage war on their people and on other countries. We need to set up the framework today to prevent catastrophe in the future.

We have discussed the implications across multiple sectors, including the inevitability of job losses, and we need to protect people, not jobs, firms, or industries. Whether the wealth generated by lower labour costs and higher productivity is shared equitably between workers and corporations is a question society must address sooner rather than later.

What we do not want is a return to serfdom, with the humanoid robot manufacturers getting ever more wealthy, and financial inequality increasing. We would like to see people able to pursue their dreams in art, music, and community-led initiatives. We need to start thinking about this now – and will look at more detail in a later chapter.

The wider global impacts are profound. Businesses in the developed world will reconsider moving production and service industries from high-cost labour markets to low-cost labour markets. Countries will likely bring back jobs they have outsourced to Vietnam, the Philippines, India, and China. The economics of humanoid robots will win over politics, hands down.

Predicting the next 20 years is very uncertain. There will be a major fight about control. Billionaires and owners of these machines will want control, and with the sheer number of robots they can manufacture, control sets up a confrontation much larger than using AI in service industries. Politicians will not want to cede control either and will be loath to give up their jobs, their influence and power to billionaires and machines. Giving control to machines (humanoid robots) is a step too far for most people and would be a folly of epic proportions. What is your perspective?

For Ada in the global south, the jobs and prosperity politicians and economists say will come with industrialisation simply won't happen.

And Tracey will be wondering where all the jobs went. She will see her job in the services sector change. She will also see the rise of humanoid robots. With an abundance of cheaper, higher-quality goods and soaring production of goods, GDP will rise at rates never previously experienced.

Conclusion of Humanoid Robots:

- They are coming. Within 5 years.
- By 2045, there will be no job that a machine cannot do.
- Humanoid robots stop drudgery.
- The risk of warfare is very high.
- Humans will not want to cede control to machines nor to the multi-billionaires who build and control them.
- The sector is worth tens of trillions. The fight for control will be bloody and protracted.

9 AI for Transportation

Introduction to AI for Transportation:

- AI enables the removal of the most unreliable component of the transport system – the human driver.
- What are the remaining technical and societal barriers to full deployment?
- What are the economics?
- The end of private car ownership for most people – what replaces it, and what happens to all that parking?
- Millions of driving jobs – taxis, rideshare, and trucking – on a clock that's already ticking.
- Beyond cars – aircraft, drones, and 3D printing as part of the autonomous transport revolution.

In Part 1 of Optimistic Storm, we examined the disruption caused by self-driving/robotic vehicles, which I prefer to call Personal Public Transport. Transport is a sector where real-world AI will accelerate industry disruption. These robotic vehicles are ushering in a new era of personal public transportation. Public transportation is successful when it meets several key criteria.[81], [82] [83] Multiple other factors drive the decision to use individual ownership rather than public transport.

- Operates 24/7.
- Cheap.
- Reliable.
- Door-to-door or less than 400 metres away.
- Safe.

Whenever it fails to meet these criteria, or cultural norms such as using bicycles or motorcycles, footpaths, or free parking at work, people will avoid public transport and drive a car. Autonomous vehicles meet those criteria. Waymo has over 3,700 robotaxis in service, provides 500,000 paid rides per week and has logged 200 million fully autonomous miles. Waymo is just one company. Robotaxis operate in over 60 cities worldwide. To differentiate from public transport, personal public transport needs to be safer (as it is only you in the vehicle) providing door-to-door service. The number of rides in autonomous cars is now higher than in taxis and ridesharing in some cities in China and in San Francisco. It demonstrates how quickly the technology has been accepted. Deployment of personal public transport will start slowly but accelerate. The USA, UK, Netherlands and other jurisdictions are beginning to implement appropriate regulations. Initial focus will be dedicated to autonomous vehicles, but it may be possible to use humanoid robots to drive vehicles instead of humans or change of the complete vehicle, in order to speed up the transition to autonomous driving using existing vehicles.

9.1 Improvements in Driving AI

AI has enabled companies to leapfrog others in the industry. In 2022, Tesla notably stated that AI could replace 300,000 lines of code for their autonomous self-driving software, known as Full Self-Driving (FSD). Tesla had spent a decade developing a code base for FSD but in 2022 declared they would change their traditional software to end-to-end neural nets. It was not a trivial task to replace that code with AI. There are important measures for autonomous vehicles. They are quite different measures, although media will often confuse them. These measures pertain to:

- Crash rates. How many km between accidents – accidents that involve either single or multiple vehicle collisions? Tesla reports a crash rate of one per 6.26 million miles driven with their FSD/Autopilot engaged, according to their Vehicle Safety Report, versus the USA rate of 484,000 miles between crashes.
- Fatality rates. Tesla vehicles are reported as 5.6 fatalities per billion miles driven, compared to the national average of 2.8 billion miles, but that may be a function of the drivers!

- Interventions. How often does a human need to intervene if the car cannot determine the next step or a critical event occurs that the AI cannot resolve, requiring a human to take over?
- Driver behaviour. The highest rate of accidents was Tesla owners with 26 accidents per 1,000 drivers, versus RAM drivers of 23 and Pontiac drivers at 10.

Except in a few exceptional events, the robotic vehicle responds better than a human. If there is a snowstorm and it is unsafe for driving, then it does not matter whether driven by a human or autonomously. In 2023, Tesla expected more than 15,000km (10,000 miles) between critical interventions, or roughly one year's worth of human driving before the average person would need to touch the wheel. Their target in 2024 was to outperform the US national accident rate. A factor of over 12 has exceeded that measure. In June 2025, they launched a pilot autonomous ride-hail service in Austin, Texas, and plan to operate in multiple cities across the USA by the end of 2025. That timetable slipped, for a change, and they now predict the timing to be then end of 2026. Federal government regulations are changing to enable national registration, eliminating the need for local and state regulations.

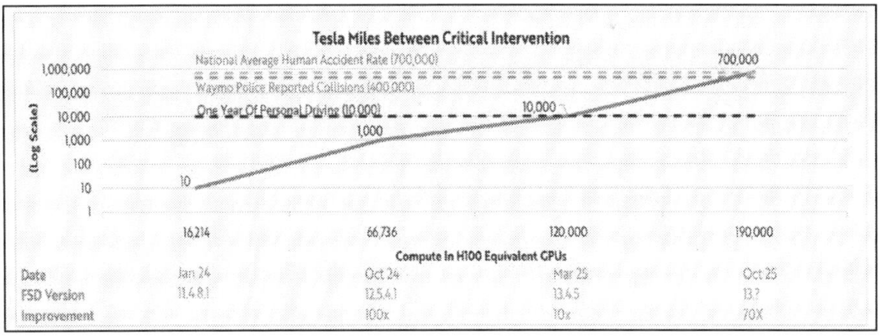

Figure 24 Tesla miles between critical interventions.

Figure 24 from Ark [39] showed the remarkable increase in miles between critical interventions. In January 2024, the miles between interventions were just 10. In six months, the number had increased tenfold to 1,000, and six months later to 10,000, and is expected to reach 700,000 by the end of 2025. It is easy to criticise Tesla for broken promises. It is a salutary example of the disruption principle, illustrating that the investment required to achieve the output was many times greater than companies initially thought. There are dozens of companies that started and failed. Ford, GM and Apple failed. The Volkswagen Group had, at one stage, over 10,000 developers, millions of lines of code, and spent over $20 billion to develop their system. What's interesting is the remarkable improvement in AI that has enabled Tesla, Waymo, Zoox, and various Chinese companies to get to the finish line. Now the race will be to manufacture sufficient numbers to dominate the market.

Waymo/Google had pinned its plan on detailed mapping, but it has also moved to neural nets for controlling its vehicles. Waymo's commercial Robotaxi service has surpassed 400,000 miles since its last police-reported collision, and the increasing use of AI has contributed to that rise. In June 2025, Waymo confirmed [84] [85] performance of deep learning models for autonomous driving scales with model size, dataset size, and training compute. Which is unsurprising as LLM models do. Using over 500,000 hours of data, they showed that more data provides better autonomous driving. At that same time, Zoox (an Amazon company of about 3,000 people) was driving around the UK and Europe on an AI system. Zoox has just announced a $10,000-per-year manufacturing facility and plans to launch across US cities in 2025 [86].

9.2 Cusp of AI Autonomous Driving

The adoption of AI enables other companies to catch up and make up for lost time. In February 2025 BYD announced autonomous driving for all of their vehicles based on AI [87] and claimed even their $10,000 car was capable of taking a 600-mile joyride on Chinese highways without driver intervention. Before their announcement, they had not even been considered a front runner, but they have driven millions of miles and state they have a disengagement rate better than Waymo/Google.

At what stage do we determine whether this technology will move from beta to production? Waymo and others have had autonomous driving and ride-hail systems for nearly 8 years, with millions of kilometres across 60 cities. My conclusion is that Tesla is five years ahead of others. Others will make up lost ground, but what are the chances and the timeframe? When people say Waymo's approach can outcompete Tesla's, it is clear they are highly unrealistic. No aspect of Waymo's business beats Tesla's model. Waymo has about 3,700 cars on the road and plans to increase that number. This is not enough vehicles. They don't win on strategy, they don't win on initial cost, they don't win on operating cost, they don't win on energy efficiency, they don't win on personnel efficiency, they don't win on technological simplicity vs. complexity, they don't win on real world driving capability, they don't win on real world data collection of edge and corner cases, and they don't win on organisational talent/capability. It is pretty clear they will lose until they change strategy. If Waymo joined up with Kia/Hyundai /SAIC/Geely, they might be able to launch a low-cost EV in 2 to 3 years. That would be 2028. Tesla can use millions of its current fleet, and it is building a production line to produce over 2 million specific "robocab" autonomous cars each year as well as its 2 million existing models. That means over 20 million cars in service by 2030. Others will follow. By 2035, there will be multiple manufacturers of autonomous vehicles, and a few niche vehicles but the industry as we know it now will be gone.

9.3 End of private ownership

The consequences will be the end of private car ownership for most people in the Western world. It also implies a potential reduction in the number of vehicles from 1.6 billion to approximately 200 million cars. Users will benefit as their transport costs will fall by 50%. RACQ reports [88] the average Brisbane, Australia, family household spends over $AU19,600 per year on transport with 2.1 cars per household. Private Public Transport models (Autonomous ride-hail) show a 50% reduction in transportation costs if they switch the second car to an autonomous rideshare and a 70% reduction if they fully switch. There will continue to be vehicle ownership for towing caravans (720,000 caravans in Australia), fleet ownership, and collectors, but for the vast majority of people, a robotaxi will be the preferred choice.

9.4 Transport Job Losses

It is probable that the deployment of autonomous transportation will roll out in stages. The rollout will first target taxi and rideshare services. The second phase will target individuals who cannot afford taxis or ridesharing services due to cost and those who cannot drive for health or age reasons. As transportation costs reduce by 50% or more, the rate of increase will accelerate. Autonomous vehicles will augment public transport. Taxi and rideshare drivers will lose their jobs over the next three years. Numbers are approximate. The USA has about 161 million workers in total, so the paid drivers segment is not a huge percentage. The losers from this change include:

- Taxi drivers in USA = 370,000.
- Rideshare drivers USA = 750,000 Uber/Lyft drivers full-time equivalents. [89]. Part time equates to ~2m Lyft drivers and ~1.5 million Uber/Lyft.
- Uber has about 24% of the food delivery services.
- DoorDash has over 60% of the market.
- Male Uber drivers work about 18 hours per week, and females about 13 hours.
- The churn rate is high, with some attrition rates of over 65% - partly as the drivers get more work in other industries, and they use ride-share driving as a part-time job.

This is just the tip of the iceberg of job losses. The trucking industry employs 3.6 million in the USA and over 30 million in China. It won't be long before trucking becomes autonomous due to the cost savings in that industry. Driving will be a job that will end. The only question is how fast. By 2028, expect ridesharing in most countries to be 100% autonomous as Uber with its 6.4 million drivers in 10,00 cities across 72 countries change to autonomous vehicles. When will all new vehicles be autonomous? By 2030? When will the majority of driving be autonomous? Within a decade.

9.5 Trucking Job Losses

The trucking sector is a major contributor to global emissions, and in Optimistic Part 1 we looked at the 18% of transport emissions and saw that a major part of these emissions is from just 5% of vehicles on the road. What is the impact of AI for the trucking sector in Australia? The transport sector has used machine learning in the distribution process, where food is grown around the world and shipped to huge distribution centres. When a consumer buys a tub of tomatoes in a supermarket in Munich, the call goes out to the warehouse in Poland, and a replacement tub is transported overnight by logistics truck, and the shelf is restocked before customers show up in the morning. That process will continue to be refined by AI. In Australia, the trucking industry employs about 250,000 truck drivers, of all truck sizes. There is a driver shortage of about 28,000. The six-wheeler truck fleet has about 80,000 drivers. In context of jobs, it is twice the number employed in mining. [1] In 2021, the annual operating cost of drivers was 40% of the costs. Fuel was 32%.

- Driver labour = $195,000
- Fuel-related + $129,000
- Vehicle = $74,000
- Maintenance and tyres = $67,000
- Registration= $6,000
- Miscellaneous costs of $15,000.
- Total Cost is $0.5 million.

A switch to autonomous electric trucks, such as the Tesla Semi, would be appealing, with 25% of costs gone, but the fleet operator would struggle to raise the capital needed to replace the fleet. Whereas, if humanoid robots with a driving app cost $50,000, the economics are compelling with 40% of the cost gone. Combining driving and electric, the cost savings are over 65%. Is it possible? Tesla, Waymo, or PonyAI could transfer their self-driving vehicle application to a humanoid robot, and create a market to replace all 60 million truck drivers globally. For every driver replaced with a humanoid robot, owners get their money back in 3 months. This application of humanoid robots is complex, at an early stage, and may never happen, but the economics are compelling. A university in South Korea took existing flight simulators, AI, and a simple robot that could sit in an aircraft cockpit seat and, within 6 months, had the robo-pilot flying the simulator better than any of their most skilled and experienced pilots. [90] Will society accept robots flying planes? Even if they demonstrated less risk? What about trucks? Advances in AI suggest this robotic truck driver would be entirely feasible. Improvement in AI over the next decade will make it a trivial task. Trucking contributes to 16% or more or road accidents. Lower cost of transport may bring savings to fleet operators and food giants. Safer driving brings benefits to society as a whole.

9.6 Land Use Transformation

AI for transportation with autonomous vehicles has an enormous impact across multiple land uses, these being:

- Agriculture. The number one crop grown in the USA is corn.
- Over 40% is converted to ethanol and mixed with gasoline for transport. Literally we are burning food in our cars.
- Over 40% of agricultural land in the USA will be freed up. Brazil, Europe and Australia have smaller quantities – see the section on land use.
- 60% of oil produced is for transportation. All those oil wells will be gone.
- Along with the impacts in food production with precision fermentation, 70% of the agricultural land area will be retired and available for regreening.
- Car parking. Los Angeles has an area the size of San Francisco in parking lots. Some cities are 1/3rd parking areas. With robotic vehicles, all that land could be put back to people's use. Cities become cities for people, not cars.
- On-street parking becomes bike lanes.
- Garages in homes. Most houses have a 2-car garage, which can account for more than 10% of the house's area. That garage can be converted into additional living space or space for rent. Society talks about tiny houses – most garages are much bigger than tiny houses.
- Financial implications including the loss of the mechanical repair industry. The number of mechanics in Australia is over 60,00.
- Urban air quality benefits.

These impacts are discussed in greater depth in Part 1 of Optimistic Storm. None of the 24 or so fossil fuel-producing countries will be immune. That includes countries such as Singapore, which do not have fossil fuel reserves but are hubs for oil refining and trade.

9.7 Automating Aircraft

Computer systems fly most modern aircraft. Some aircraft, particularly military aircraft, have to be flown by software as the planes are at the edge of physics and pilots cannot control them. Most commercial planes operate on automatic systems most of the time, and pilots only take control during approach or take-off. Researchers at the South Korean Advanced Institute of Science and Technology (KAIST) developed a humanoid robot that could "fly" a flight simulator better than their best Korean airline pilots. They designed it to fit in a pilot's seat as they wanted something that would not alter the plane in any way. They called their robot Pibot. They also tested their Pibot on driving cars and tanks as well as aeroplanes, and [90] Pibot could drive that vehicle with no vehicle modification. If a Pibot flight pilot robot finds a safer way to land in crosswinds, that knowledge would be passed to every single flying Pibot robot, every robot globally, and potentially immediately. No 2-month flight simulator, no supervised flying and 330,000 pilots will not need updated training and certification. Will we see this technology in commercial aircraft soon? Unlikely, but given that at least 70% of aircraft crashes are due to "human error" and only 20% to mechanical failure, expect some changes. A Pibot sitting in one seat and a human in the other seat might be the first step.

9.8 Saving money, Saving Lives

Expect to see a "march of nines" where the next goal is ten times safer each progression. E.g. 99.9% is ten times less than 99%, 99.99% is ten times less than 99.9%. Autonomous vehicles are now safer than 99.9% of drivers. At the next nine, 99.99%, the accident rate could be 10 times less. At 99.999, it is over 100 times less. We know that 70% of road fatalities occur between half an hour before and after sunrise and sunset. Furthermore, most accidents occur within 40 km (25 miles) of the trip's origin and destination. These peak accidents are when drivers are commuting, driving on "automatic", and not paying attention. Autonomous driving will be significantly safer. Improvements in AI appear to be the main reason, as neither humans nor hardware in the vehicles has changed.

My optimism about moving to autonomous driving and, in particular, personal public transport is driven by society's view that private cars are a human right. In Part 1, we noted that Australian road fatalities are about 1,400 per year, with 62,000 people hospitalised. These occupy 18% of hospital beds, with health costs exceeding $AU30 billion a year. That number is close to 20% of the health budget and 2% of GDP. Australia's rate of 4.54 deaths per 100,000 population was ranked 18th out of the 31 nations and just above the OECD average. Nations with the lowest rates were Norway (2.14), Sweden (2.17) and Iceland (2.39), whereas the USA is three times higher at 12.8. India is one of the most dangerous countries at over 16. When autonomous vehicles are 99.9% safer than humans at driving, the road toll would be 140 people in Australia. At

99.99%, the road toll would be just 1 person, with 60 people hospitalised. Savings of $30 billion from health budgets are enormous. Saving lives; priceless. We will look back on the era of driving and road injuries the same way we look at smoking or the use of asbestos. A needless loss of life with devastating ramifications. Autonomous driving today is the worst it will ever be, and we can be confident that by the time it reaches the UK or Australia in 2027 or 2028, it will be a "9" or two 9s better, thanks to prior deployment in the USA and China and the continued improvement in AI.

9.9 Drones

We all have some experience with or exposure to drones. Whether as enthusiasts for photography, amateur fishers, viewers of magical displays, recipients of Amazon goods, or consumers of videos from Ukraine, Iran, or other conflict zones.

9.9.1 Commercial Use

Until recently drones were controlled by operators on the ground, but increasingly are controlled by AI. They range from tiny to 4-seater eVTOL planes. Both commercial and military sectors are booming. Collaboration and technological advancements are enhancing drone capabilities, making small drones indispensable in various applications. DJI is generally regarded as the leader in the consumer and commercial drone market. Other prominent companies include Parrot, AeroVironment, Autel Robotics, and EHang. For military drones, companies include AeroVironment, Northrop Grumman, and Saab. Almost anyone can manufacture a drone using 3D printing and AI control systems today.

The commercial drone market size has grown strongly in recent years. It grew 9.5% in 2024, and is expected to be $3 billion in 2029 at a 5.8% CAGR. Expanding use cases, data analytics and AI integration are powering the uptake. Some suggest we see the rise of the drone-as-a-service (DaaS) model.[91] McKinsey [92] stated companies could save 40% of their last-mile delivery costs using drones, with a 15%–20% increase in their profit margin and a 15-20% decrease in their product or service prices.

9.9.2 Agriculture Use

The agriculture sector has adopted drone technology rapidly. Part of the reasons is the labour is scarce and most people do not want to sit for hours in a tractor. Secondly, drones enable multiple passes over cropland without leaving any tracks, so can be used in wet soil conditions. And most importantly, they are not near airports or people, so the regulations and safety requirements are less onerous. Drones monitor crops and livestock, provide water management, pest management, and provide farmers with valuable data.

9.9.3 Killer Drones

Rising geopolitical tensions and the need for advanced surveillance and strike capabilities drive the military drone market. The sector is expanding rapidly, and a recent article examines how both Ukraine and Russia are utilising lethal drones. It highlights drones with a 300km range that cost less than $10,000. These perform tasks that a $3-$5 million rocket would typically accomplish [1]. A $300 drone can spot a person, follow them, and even kill them. Six drone jet skis sank a Russian destroyer [1]. In Operation Spiderweb, 117 Ukrainian drones targeted Russian airbases deep inside the country on 1 June 2025, aiming at the Kremlin's nuclear-capable long-range bombers. These $600 drones destroyed over 20 aircraft worth hundreds of millions. Deployed from the tops of trucks, the drones employed "terminal guidance" software to fly autonomously in the final mile if Russian jamming systems cut off their connection to the pilots. Note that this technology was 18 months old. A first mission for the Jiu Tian, a Chinese mothership drone, is reportedly imminent; it is said to fly at an altitude of 15 km, with a range of 6,400 km, carrying up to 6 tonnes of ammunition and up to 100 autonomous drones. Ukraine is also developing more advanced drones, where the targets are not just tanks, planes, railway hubs, and critical infrastructure, but where the primary aim is to kill people. The CEO of a drone manufacturer in Ukraine[93] says the risk is extremely high that technology developed for warfare will be used outside of the war zone. All these checks that we go through in airports are completely useless already, so we are just wasting our time," he says. "You don't need to bring a bomb to blow up the plane. You can just wait outside with the drone and wait for the plane, for the prime minister."

The military drone market was valued at $11.2 billion in 2025 and is expected to double in ten years at 8% per year to 2034. Autonomous drone technology provides services ranging from reconnaissance to combat operations. [94] As with killer humanoid robots, we don't know where this will end up. The USA is investing heavily in R&D as they saw how Ukraine destroyed planes. Countries with poor human rights records will use robot drones for surveillance of their own people.

9.10 3D Printing

3D printing is a specialised form of robotics that has dramatically reduced both the part count and the product development timeline. Manufacturers hold less inventory and save on tooling costs. In design and architecture, 3D printed models have become standard for identifying problems before construction. In healthcare, 3D printing is making novel surgeries possible with customised guides, tools, and prosthetics and implants. 3D printing for joint and organ parts reduces the time spent in operating rooms, saving more than $12.5 billion. In food, 3d printers are producing meat and protein products that mimic real food. Ark Invest explores this in their Big Ideas [49, p. 2] and estimates that robots and 3D printing could scale at a ~80% annual rate over the next eight years.

By using sensors and machine learning algorithms to control the printing process, researchers have reduced print errors by approximately 30%, strengthened products by up to 2 times, and reduced material waste by up to 40%. As consumers, we won't notice much of these changes; they will occur behind the scenes. Humanoid robots will accelerate the use of 3D printing.

9.11Conclusion Transportation

AI is speeding up autonomous vehicles, and we see that there are both first-order and second-order impacts being:

- Transport becomes cheaper, cleaner, safer, and more reliable.
- Reduction in pollutants improves people's health.
- Large increase in GDP.
- Cities can be changed for people, not cars

- Parking lots are up to 30% of cities and that space is given back to people.
- AI for knowledge workers combined with safer and cheaper commuting will mean workers can live further away from city centres, taking advantage of lower-cost suburban housing.
- Roads become bicycle-friendly with bike lanes and safer cars.
- Electric bicycles turn every road flat. Short distances and safer roads will make bicycles a reliable mode of transportation.

Conclusion to Transportation:

- Most vehicles within a decade will be autonomous.
- Compelling economics will remove the driver to stop road fatalities.
- We will look back and wonder why we tolerated death and destruction and wonder why millions had to die.
- Cities will reclaim vast amounts of land currently used for parking and roads.
- Millions of driving-related jobs will disappear. Will transport-as-a-service create new ones?
- The transport disruption cascades directly into agriculture, energy, and city design – nothing here happens in isolation.

---

10 Agriculture Disruption By AI

How will Agriculture be disrupted?

- There are 2 types of agriculture globally. Small and Large
- What are the changes to agriculture over the next 2 decades?
- What are the major changes expected?
- How human land use has already reshaped most of the planet – and what that means for what comes next.
- Precision agriculture and robotics – AI-driven sensors, drones, and autonomous machinery on the farm.
- Soil health and regenerative farming – can AI help repair what industrial agriculture has degraded?

Human land-use activities are a terrible force, affecting everyone. [95]. We have extensively modified the Earth's land surface, altered ecosystem structure and functioning, and diminished ecosystems' ability to continue providing valuable resources such as food, freshwater, and forest products, and services such as regulation of climate, air quality, water quality, and soil quality.

Humans have altered the world in ways no previous species has, altering three-quarters of the land-based environment and about two-thirds of the marine environment. The tropics are most severely impacted by the loss of intact ecosystems. They were once home to the highest levels of biodiversity on the planet. For example, 100 million ha of tropical forest were lost from 1980 to 2000. Where?

- Cattle ranching in Latin America (about 42 million ha).

- Plantations in South-East Asia (about 7.5 million ha). About 80% is used for palm oil, mostly in food, cosmetics, cleaning products and fuel.

We often use agriculture and food interchangeably, but in reality, much of what we call agriculture is used to produce energy or animal protein. Very little is used to produce what we think of as food. As we will explore in the chapter on food, these vast areas are among the reasons agriculture has been so destructive.

10.1 Land Use

Most of us are ignorant of how much land Earth has, how much is used for food production, and how much is used for infrastructure. Before we examine agriculture per se, let's first explore how much land we use for various purposes. A 2025 UN Report says humanity has altered 70% of the Earth's land, putting the planet on a "Crisis Footing". Globally, agriculture is the biggest land degrader, and livestock are the worst. Half of the world's ice- and desert-free land is used for agriculture. Most of this is for raising livestock – the land requirements for meat and dairy production are equivalent to an area the size of the Americas, spanning from Alaska to Tierra del Fuego.[96] The facts are:

- Human habitation occupies 1% of the planet's surface.
- Crops occupy 12%.
- Livestock grazing occupies 28%.
- 2/3^{rds} of the land used for pasture is not suitable for crops.

The world has large areas with minimal habitation, so let's take a look at three developed countries, Australia, the USA and the UK, to see how they fit these global numbers.

10.1.1 Australia

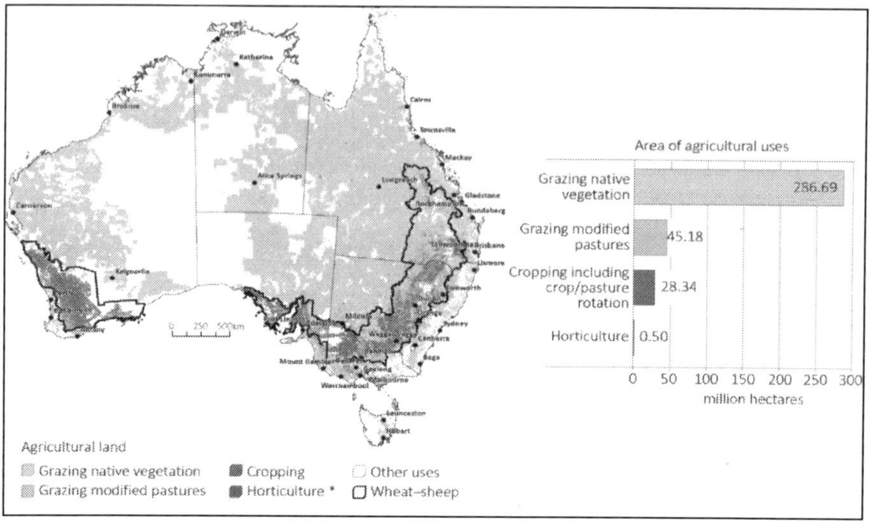

Figure 25 Australia's land use with 290m ha of grazing native vegetation

Figure 25 shows that Australia is an "empty" land, with the top land use numbers [60] [61] being nature conservation (9.5%), managed resource protection (18%), other minimal use (14%), grazing natural vegetation (42%), and grazing modified pastures (6.1%). These add up to a majority at 89.6%. Urban land use is just 0.62%, and mining and water are 0.09%. Removing animals from grazing natural and modified land would turn 50% of Australia back to managed or nature resource protection.

- The focus on urban use is misplaced.
- Agriculture is the primary cause of habitat loss.

Unmitigated land clearing for agriculture is, and continues to be, a disaster for habitat loss. Australia has put this land into the hands of only 132,000 farmers, but farm sizes are highly skewed. Ten per cent of broadacre farms produce around half of the total output, while the smallest 50% of farms produce around 10% of the total output. Beef farms are even more skewed, with 10% producing 60% of the output. Only 10% of Australian farms are cropping farms.

10.1.2 UK

Is the UK crowded? [97] As a tourist, it appears overwhelmingly built up. That is an illusion because it is only a little more built up than Australia. For the UK:

- 8.7% of the land is developed.
- 91.1% is undeveloped.

Agriculture remains the largest land use, accounting for 63.1% of land, with over 209,000 holdings occupying about 17 million ha. As in Australia, farm size is highly skewed: the average is 80 hectares, but over half are less than 20 hectares. Forestry, open land and water make up the balance. Over 72% of the agricultural land is pasture, with 26% cropland. Land for farming is decreasing, with some moving into infrastructure, but also into uses including golf courses and woodland. If you have ever travelled to England and marvelled at their many residential gardens, it stands at 4.9%. Of England's developed land, residential land is about 1.3%, and industrial and commercial land is about 0.4%. Driving through industrial estates is deceptive. History tells us England was a wooded land from Roman times through the 16th and 17th centuries, when woodland was cut to provide timber for boats and charcoal for the iron industry. More woodland has been cut down during the 20th century than for centuries beforehand.

10.1.3 USA

The USA has a land area of 941 million hectares (2.26 billion acres) and similar land-use patterns, with 46% used for agriculture. The breakdown of land use is as follows:

- Grassland pasture and rangeland 29%.
- Cropland 17%.
- Forest-use 28%.
- Parks and wildlife areas 14%.
- Miscellaneous other uses (such as wetlands, tundra, and unproductive woodlands) at 9%.
- Urban land just 3%.

10.1.4 Land Use Globally

Even in populous India [98] agriculture is the major land use at over 51%, forestry at 21% and 4% under pasture. The built-up areas are about 7% (including buildings, roads, rail and infrastructure) and uncultivated land (5%).

While our perception may be that cities and built-up areas are the dominant land use, the reality is that agriculture is. Some may be in cultivation, such as vast areas in the USA, much of Europe, and the UK, but in most countries, the predominant land use is grazing. Moreover, the commonly held view that cities should not encroach on rural areas because the land is too valuable is misplaced, given that we can reduce agricultural land use while increasing overall production.

Livestock grazing is incredibly inefficient. If we improved the energy efficiency of agriculture to just 1%, we could support 100 billion people on less than 10% of the land we use today. Plant growth is typically not limited by the availability of sunlight, but rather by nutrients and water. By using more energy to provide better growing conditions, we can grow much more food on much less land. Modern agriculture is an order of magnitude more productive per hectare than pre-industrial agriculture. We will explore the technology we have today to make it another two orders of magnitude more efficient.

10.2 Agricultural Subsidies

Transforming farming practices could restore billions of hectares by 2050 for less than is spent on developed-world farm subsidies. [99] Read that again. For less than current subsidies!

We won't explore in detail how agricultural subsidies managed to snare such high levels of subsidies. Agriculture was once the dominant industry, and most politicians were large landholders. Government intervention in food and fibre commodity markets began centuries ago with the classic case of farm subsidy through trade barriers such as the English Corn Laws that regulated the import and export of grain in Great Britain and Ireland. They were repealed in 1846. Modern agricultural subsidy programs in the United States began with the New Deal and the Agricultural Adjustment Act of 1933. [100] The OECD estimates that current agricultural subsidies total over $850 billion across the 54 countries it reported on.[101]

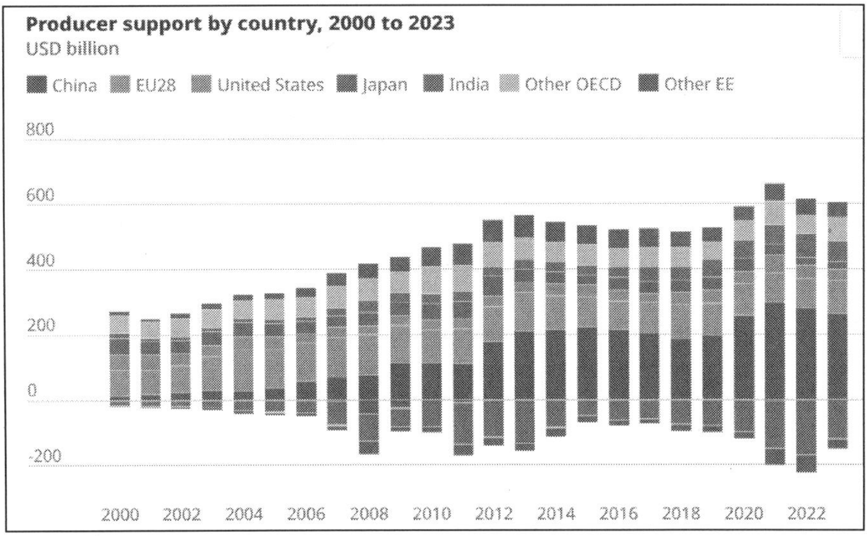

Figure 26 OECD data 2024. China is 37% and India 14%. USA is now 15% and EU 13% of the global total down from 50% in the early 2000's.

As Figure 26 shows, producer support has trebled in the past 2 decades from about $250 billion to over $600 billion and the data shows that more and more goes to the larger holdings, which is typical of all government interventions, that a few capture most of the subsidies and political and social considerations outweigh efficiency considerations in the allocation of subsidies. [102]

The political challenge will be to reallocate those subsidies. Subsidies, once they become part of society's expectations, become very difficult to stop politically.

10.3 Agriculture Emissions

Agricultural emissions are also rising, and it is not just CO_2. Methane's global warming potential (GWP) is 27–30 times greater than CO_2 over 100 years, and nitrous oxide has a GWP 273 times greater than CO_2 over that timescale. The table below underscores the mitigation benefits from reducing agricultural methane and nitrous oxide. Livestock and manure emissions account for 30% of agricultural emissions and nearly 6% of all emissions.

Sector	% Global Emissions	Million Tonnes CO2
Livestock and manure	5.8%	2,175
Agriculture Soils	4.1%	1,538
Rice Cultivation	1.3%	488
Crop Burning	3.5%	1,313
Deforestation	2.2%	825
Cropland	1.4%	525
Grass land	1.0%	375
Total Agriculture	19.3%	7,238
Total Global Emissions		37,500

Figure 27 Agriculture contribution to global greenhouse gas emissions

Here are some facts about agriculture.

- Food, agriculture and land use emit about 19%-22% of greenhouse gas emissions directly.
  - 11% or half of this is tropical deforestation and the subsequent farming practices.
  - 5% of emissions is methane from livestock. The methane from animals is more than the methane from coal and gas.
  - 3% from industrial agriculture practices.
  - 2% from rice methane.
- The wider food system emits about 34% if we include the growing, fertilisers, transport, energy and waste.
- Carbon removal on farmland is less than 0.1%.
- Even in the USA, agriculture accounted for 11% of GHG in 2020.

Compared to agriculture of 19%, cement is about 8% of global emissions, and aviation is 1.5%. Whereas in some industries, CO_2 is the majority, for agriculture, methane and nitrous oxide. Land converted to cropland or grass is a major contributor, depending on the land use following deforestation. Emissions from 1.5 cows is equivalent to a single car. Nitrous oxides from soil and manure management are critical contributors to GHG emissions. Sequestration is a small part of emissions and can't keep up with the huge ones. Agriculture is a world of two quite different structures. One structure is highly subsidised, dominated by global giants, and wields substantial political power.

The other is a number (4 billion) of very small subsistence farmers with less than 1 hectare of land.

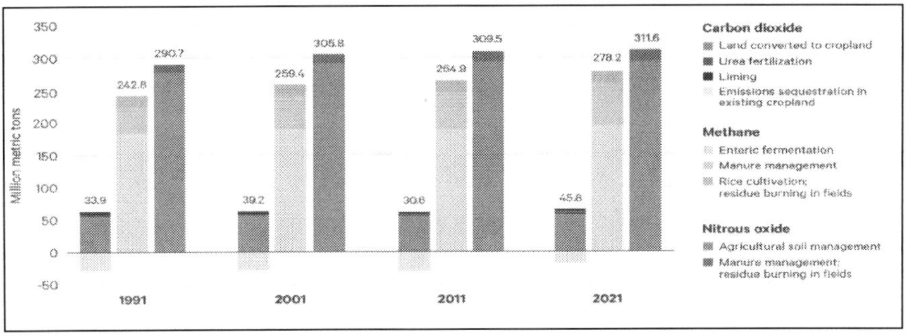

Figure 28 Agricultural emissions CO2, Methane, and Nitrous Oxides

Figure 28 breaks down CO2, CH4, and NOx over the past 30 years into their respective emissions. The actual total emissions have not changed much over those decades. CO_2 is mostly due to land converted to cropland. Methane is still 70% manure management, and NOx are 90% manure management. In fact, one could say that methane and nitrous oxides are all about how to deal with the shit better! Being an animal, that is a fact of life.

10.4 Agriculture Degradation

Agricultural activities, in particular, have been responsible for a vast majority of these land use related ecosystem consequences [103] [104] [105]. Nearly 40% of the planet's ice-free land surface is now being used for agriculture, and much of this land has replaced forests, savannas, and grasslands. [95] Clearing of tropical forests for cultivation or grazing is responsible for about 12–26% of the total emissions of carbon dioxide to the atmosphere [106] [107] and land use changes can significantly modify regional and global climate [108] [109] Furthermore, about 20–30% of the total available surface water on the planet is withdrawn for irrigation and that has grown from 0.5 billion m^3 to 4 billion m^3 [110] since 1900. Nitrogen fixation through fertiliser production and crop cultivation currently equals or even exceeds natural biotic fixation.

As we discussed in climate change, 19% of direct emissions are attributable to animal agriculture, with about 80% of agricultural land going toward animal agriculture. Our global meat consumption has risen by 70% over the past 26 years, to over 360 million tonnes. With only 26 years left until 2050, and with continued increases in meat production and consumption, climate goals will be impossible to meet. Biodiversity and deforestation goals will be much harder to achieve.

Animals are extraordinarily bad at converting food to protein. The numbers speak for themselves.

- A chicken needs 9 calories to get just 1 calorie out as chicken meat. The majority just goes to keeping the chicken alive.
- The consequence is that the chickens need 9 times the land, 9 times the water, 9 times the waste, and 9 times the pesticides and herbicides to grow the crops.
- It is literally 800% food waste in the physiology of the chicken.
- Worse, cows need 40 calories to get 1 calorie out.
- That does not account for the emissions from the industrialisation.

What is the solution? Bypass the chicken or cow altogether? We will explore this in the food chapter.

10.5 Waste

We are told our agricultural systems are efficient, but in reality, our agricultural and food systems are wasteful. Tackling the scourge of food loss and waste represents a triple win opportunity - for the climate, for food security, as well as for the sustainability of our agri-food systems, says the FAO [111]. Global food waste is over 2.5 billion tons annually, with about 14% lost from the farm to the shop, and a further 17% at the retail and consumer end. The food that is lost and wasted could feed 1.26 billion hungry people every year, as well as a further 1.1 billion people facing moderate or severe food insecurity. FAO estimates it is $400 billion down to the dump! Even those in the Western world would save hundreds of dollars each year. [111] Those numbers have not changed for the past 5 years.

Wasted food also contributes to the burden of emissions, comprising over 11% of all emissions.

The USA discards more food than any other country in the world: 30% to 40% or over 40 million tonnes. USA households contribute 43% of food waste, restaurants and food service 40%, 16% from farms and 2% from manufacturers. Recent programs in Australia to reduce waste by the duopoly supermarkets who control 70% of food sales, reduced waste by 17% or over 100,000 tonnes. That equates to 505,000 tonnes of CO_2 avoided. Environmental benefits go beyond emissions, with food waste reduction saving water, fertiliser, arable land and hundreds of millions of dollars. Of the food wasted each year, about 70% is perfectly edible. However, in most countries, tax frameworks treat food donations the same as waste disposal, discouraging businesses from donating surplus food to charities.[112] Waste not only affects consumers but also food service businesses who otherwise dump their food, and contribute to the mountains of municipal waste.

We have all the technology needed to stop that wastage and emissions, but the economics and behaviour limits how quickly we can change. A simple tax change would change economics and that will drive change.

Tracey's household budget would save more than $600 each year. As we have seen, waste is really localised. Food waste impacts all countries - even those who have little starvation per se. It drives cost of living pressures. We see people in developed countries go without food because their energy / rent costs take up all their disposable income. The UN 2030 Agenda for Sustainable Development calls for the halving of per-capita global food waste but unfortunately, reducing food waste is not sexy politically. Just like energy efficiency, politicians like to focus on new technology, and not expand the use of existing technology to gain savings. In context, emissions from all flying done around the world is dwarfed by emissions from food waste. If you like taking that holiday flight, consider reducing your household food waste. Don't spend a few dollars flying green - stop your own food waste!

10.6 Regenerative Agriculture

Regenerative agriculture is a conservation and rehabilitation approach to food and farming systems. It focuses on topsoil regeneration, increasing biodiversity, improving the water cycle, enhancing ecosystem services, supporting bio sequestration, increasing resilience to climate change, and strengthening the health and vitality of farm soil. (Wikipedia [113])

Regenerative agriculture has recently received significant attention from producers, retailers, researchers, consumers, politicians, and the mainstream media. Despite widespread interest in regenerative agriculture, no legal or regulatory definition of the term "regenerative agriculture" exists, nor has a widely accepted definition emerged in common usage. [114]. The term has only increased since 2019.

10.6.1 Top Ten Claimed Benefits

Much of the information on regenerative agriculture comes from NRDC (the Natural Resources Défense Council). They claim 3 million members and online activists with the expertise of some 700 scientists, lawyers, and other environmental specialists, and their purpose is to confront the climate crisis, protect the planet's wildlife and wild places, and to ensure the rights of all people to clean air, clean water, and healthy communities. A WEF forum also puts forward a range of benefits. [115]

Claimed Benefit	Description
Healthier Soil	Improves soil fertility and organic matter content
Reduced Erosion	Protects soil from degradation and loss
Climate Change Mitigation	Captures carbon dioxide in the soil
Increased Biodiversity	Supports a wider variety of plant and animal life
Improved Water Management	Enhances water retention and reduces runoff
Boosted Crop Yields	Leads to more productive and resilient crops
Reduced Reliance on Chemic:	Minimizes the need for pesticides and fertilizers
Increased Farm Profitability	Lowers costs and potentially creates new revenue streams
Farmer Wellbeing	Enhances farmer livelihoods and resilience
Healthier Food	May contribute to more nutritious food production

Figure 29 Claimed benefits of regenerative agriculture

10.6.2 Electrification of Agriculture

Farming is mechanised in the Western world with substantial on-farm costs of diesel and electricity. About 10% of the agriculture sector's emissions are from on-farm fuel combustion. Many farmers are investing in on-farm solar and batteries for irrigation and for dairy farm milking sheds. Horticultural farm machinery is moving to electric, but the large cropping sectors have few products. High torque and long duration of the machinery have seen much talk, and little action. Precision agriculture using drones is a burgeoning area of change and can reduce the number of trips by fossil-fuel machinery over the paddock during a crop lifecycle.

Change will be slow. Most countries have generous farm subsidies of fossil fuels and without a large price differential, we won't see farmers changing rapidly. [116]. Some suggest intermediate steps with hybrid, but heavy trucking has similar issues. A rapid rise in battery energy density will see the pace of change pick up, but expect to see tractors and combine harvesters with us over the next few decades. There is almost a much better return on investment in reducing food waste on farms, reducing fertiliser, or reducing water use than in changing a tractor from diesel to electric.

10.7 Pasture Fed Organic Beef

Some claim that pasture-fed organic beef has much lower emissions than feedlot beef. Nothing is further from the truth, and no major farm product is as environmentally damaging as organic, pasture-fed beef. George Monbiot explains this in detail in "Regenesis: Feeding the World Without Devouring the Planet" [117]. He summarises this in an X/Twitter post:

1. The most important environmental metric is the amount of land we use. Land used for extractive industries is land that cannot be used for wild ecosystems such as forests, wetlands and savannahs, on which the great majority of terrestrial species depend. Human habitation occupies 1% of the planet's surface, crops 12%, but livestock grazing occupies 28%.
2. Animals fed on pasture alone produce just 1% of the protein we consume.
3. If the grain-fed cattle in the US were instead raised on pasture, they would need 270% more land. [118]
4. In other words, even if the US felled all its forests, drained its wetlands, watered its deserts and annulled its national parks, it would still need to import most of its beef. And that would be from Amazon, which is already meeting part of the rising demand for "pasture-fed" beef.

5. The climate costs of all beef are massive: raising a kg of beef protein, on average, releases 113 times more greenhouse gases than a kg of pea protein, and 190 times more than a kg of nut protein. But pasture-raised beef is much worse: 3 or 4 times the emissions of grain-fed beef. [119]
6. Organic beef production loses twice as much nitrogen per kg of meat as conventional beef due to the longer time it takes to get to maturity. Moreover, it uses even more land than conventional pasture-fed beef, so it has a carbon opportunity cost.[120]
7. Carbon that could be stored in wild ecosystems but can't be because they've been turned into pasture.
8. Claims about carbon storage in pasture have been hugely exaggerated and tend to rely on junk science. Studies show that pasture absorbs no more than 60% of the greenhouse gases livestock produce. [121]

9. Lamb is worse than beef, but not as big a demand

10. Livestock numbers have to be very low to be compatible with a functioning ecosystem. A Knepp-style farming covering all of the UK, would only provide 75kCal meat per day.

In conclusion, changing farming practices make little difference to the 28% of the land we use for farming, and something else is needed to move the needle.

10.8 Burping Cows

As we have seen above, the biggest agricultural emissions come from cattle, both for meat and dairy and are the number 1 source of greenhouse gases worldwide. Livestock farming accounts for 14.5% of all anthropogenic greenhouse gas (GHG) emissions, emitting 7.1 gigatonnes of carbon dioxide (CO_2) equivalent per year. That includes land clearing, grazing, crops grown for feedlots or dairy, and the waste and methane from belching and manure, which account for around 44% of livestock emissions.

Each year, a single cow will belch about 100 kg of methane. Methane from cattle is shorter-lived than carbon dioxide but 28 times more potent in warming the atmosphere. With 1.6 billion cattle, that's over 4.5 billion tonnes of CO_2 e. In contrast, cement is about 8% of emissions with 1.6 billion tonnes CO_2.

What do you do? Eat less beef as some advocate? They contend it's an unsustainable diet in a world with a population expected to reach nearly 10 billion by 2050. We discussed early on that people do not easily deprive themselves. Things we can do include:

- Eat less beef - most of that beef is eaten by the billion people who consume 72% of the earth's resources.
- There is a whole class of people who aspire to eat more protein.
- Eat less beef and more chicken or other proteins (fish).
- Stop cows from burping methane.

10.8.1 Technology to Stop Cows Burping

There are solutions to stop cows from burping.

- A red seaweed, *Asparogopsis,* demonstrably reduces methane emissions with intakes of 1 to 2% per day. As an aside, the author started a business to grow seaweed for this very use.
- Breed cows that have low natural levels of methane emissions.
- Selectively breed bacteria with lower emissions.

10.8.2 Bypass the Cow

Bypass the cow for the production of products using precision fermentation - see the chapter on food.

- Casein and fat, sugars, and reduce dairy herd by 90%.
- Ground beef .
- Stop feeding pets with animals. Over 20% of all pet food is from other animals.

10.9 Agriculture for Biofuel

We discuss in detail in Part 3, in the chapter on E-Fuel and biofuels, with the conclusion that the use of biofuel should only be for difficult-to-decarbonise sectors such as long-distance shipping. Using biofuel for light vehicle use or burning food in a combustion engine has created an environmental disaster. We will summarise that discussion, because of the huge impact on land use and land clearing of tropical rainforests.

IEA states biofuels play a vital role in decarbonising transport [122] by providing a low-carbon solution for existing technologies, such as light-duty vehicles in the near term and heavy-duty trucks, ships and aircraft with few alternative and cost-effective solutions in the long term. Their thinking is not true: batteries will power both light and heavy-duty vehicles. The whole biofuel and E10 petrol and diesel blend arguments are driven by the biofuel industry and industry lobby groups, make no climate-mitigation sense at all, and have led to the worst government subsidies. Ther is conflicting analysis. A 2019 study [127] found that corn ethanol's emissions profile was 40% lower than gasoline's. Whereas Lark's 2022 review [127] was the exact opposite, stating it was at least 24% more intensive. They put this down to the Renewable Fuel Standard, which increased corn prices by 30%, prices of other crops by 20%, which, in turn, expanded US corn cultivation by 2.8 million ha (8.7%) and total cropland by 2.1 million ha (2.4%) in the years following policy enactment (2008 to 2016). These changes increased annual nationwide fertiliser use by 3 to 8%, increased water-quality degradants by 3 to 5%, and caused sufficient domestic land-use change emissions that the carbon intensity of corn ethanol produced under the RFS is no less than that of gasoline and is likely at least 24% higher.

Of course, the ethanol producers said this was cherry-picking and poor science, and continued to claim that E10 can "reduce emissions by 5%." Biofuel demand in 2022 reached a record high of 4.3 EJ (170 billion litres), surpassing the 2019 level, prior to the COVID-19 pandemic. Most biofuel production currently uses so-called conventional feedstocks, such as sugar cane, corn and soybeans. The scale of growing crops for ethanol production is staggering. Over 105 billion litres of ethanol are produced annually, and the major producer is the USA (53%, 46 billion litres) followed by Brazil (28%, 8 billion litres). The EU, India and China are about 5% each. In the USA, total corn production is about 15 billion bushels, with about 40% used to produce ethanol. The land for corn in the USA is 15 million ha (36 million acres), and at 40% for ethanol, that's about 6 million ha. In context, Oklahoma's total farmland is only 5 million ha.

These numbers show how the entrenched farm subsidy lobby wants to stop electric vehicles. The livelihood of states is at risk. When consumers change to electric vehicles there are no more E10 or diesel additives, no oil, no biofuel, no subsidies and ultimately no farmers.

The policy failure is that ethanol is added to petrol not just in the USA, but exported to Europe and used in fuel there. One could get extremely cynical about this whole subsidised, politicised area. Or just accept that the only way to stop this is to electrify transportation. That stops the subsidies. The only potential use for future use of biofuels would be for long-distance shipping and aviation, for which batteries would be difficult. Even then, using only waste crop residue is the only sound emissions strategy. Are we likely to see that in 20 years? No. The only bright side is that the industry will pivot away from road transport to aviation and shipping, where biofuels are much more appropriate than using H_2.

10.10 Ecological Disaster

The ecological disaster that humans have created is mind-blowing, and most people have little understanding of what human civilisation has done. We see European cities with magnificent cathedrals and museums, but won't swim in the Seine or Thames without a bad case of dysentery. We see, or don't see, animals or plants in intensive agricultural areas. Intensive agriculture is hostile to most species. We touched on this in the impacts of climate change in Part 1, but we will explore some of the impacts of agriculture on our environment. These impacts are separate to the impacts of climate change.

What goes on behind the scenes of intensive agriculture is hidden. The effect of humans on global biodiversity first became significant as modern *Homo sapiens* migrated from Africa to occupy the other continents, and we lost mammoths, ground sloths, giant kangaroos and others we ate. As we moved to settle the islands, we chomped through moas in NZ and at least 1,000 bird species, which is 10% of all birds. Extinction rates in the recent past have been running 100 or more times faster than in pre-human times, and the pace of extinction has accelerated over the last few centuries. If this continues, the loss of species will soon amount to a large fraction of all species on the planet. [123]

10.10.1 Extinctions

We might have read or seen the destruction of forests - in the Amazon, in the USA, or Australia. From the IUCN red list since 1500 AD over 711 vertebrates including 181 birds, 113 mammals and 171 amphibians are gone. [124] The average abundance of native species in most major land-based habitats has fallen by at least 20%, mostly since 1900. More than 40% of amphibian species, almost 33% of reef-forming corals and more than a third of all marine mammals are threatened. The picture is less clear for insect species, but available evidence supports a tentative estimate of 10% being threatened. The best estimate now is that approximately one million of the world's species are currently threatened with extinction. And that is borne out by the increasing number of species on the threatened lists, the continuing destruction of habitat, and the impacts of climate change.

10.10.2 Loss of Habitat

Humans have altered the world in ways no previous species has. Humans have significantly altered three-quarters of the land-based environment and about two thirds of the marine environment. The tropics are most severely changed with losses of intact ecosystems. They used to be home to the highest levels of biodiversity on the planet. For example, 100 million ha of tropical forest were lost from 1980 to 2000, resulting mainly from cattle ranching in Latin America (about 42 million ha) and plantations in South-East Asia (about 7.5 million ha, of which 80% is for palm oil, used mostly in food, cosmetics, cleaning products and fuel) among others.

Global population doubled from 3.7 to 7.6 billion since 1970, along with an increase of GDP per capita of four times. These distant consumers shifted the environmental burden of consumption and production across regions. Urban areas doubled. A third of marine areas are un-sustainably fished. More than a third of the land area is now farmed and over 75% of freshwater devoted to crops or livestock.

As well as urban areas doubling, one third of the world's nature reserves are under threat to humans [125] and as the Conversation points out since the Yellowstone National Park became the world's first protected area 152 years ago, nations have created more than 200,000 terrestrial nature reserves. Together they cover more than 20 million km^2 or almost 15% of the planet's land surface – an area bigger than South America. Almost three-quarters of countries have at least 50% of their protected land under intense human pressure, and more than a third of these are under threat from roads, mines and other human activities. No country is exempt.

Scientists agree that we must protect at least 30% of lands, fresh waters and oceans by 2030 to have any chance of stemming the dual crises of biodiversity loss and climate change facing our planet. Currently, only about 17% of land and 8% of marine areas are currently under some form of protection. The "30 by 30" initiative in the Kunming-Montreal Global Biodiversity Framework was adopted in December 2022 [126, p. 30] and has over 100 countries signed up. The 30 by 30 is for one third (30%) by "2030". Some, like the EU, have agreed to 20% by 2030. But there is criticism. More affluent nations would have the ability to select vast natural parks and classify them as protected areas, without any additional benefit for flora and fauna. It's a good start, but as they say, the devil is in the details. We often think of these as federal issues, but it is local in most cases. E.g. On the Sunshine Coast, the local council wants to establish a "blue heart" [127] for restoration of old cane farmland. Developers want a water park and tourist amenities. Who wins? Not nature.

10.10.3 Water

Earth should have been called Ocean as 70% of Earth is covered by water. The ocean is neither drinkable nor useful. Only 2.5% is fresh with 1% accessible. Water inequality is severe. The drinking water shortage is so severe today that by 2030, nearly half the world's population or between 2.8 to 3.4 billion people may be exposed to severe water stress even under median scenarios [128]. They will lack safely managed drinking water. Over 700 million people won't even have basic water services. Shortages are driven by demand [129] and follow similar patterns across the world. [130] Expanding global population, economic developments, and agricultural needs all require water and infrastructure. Water use has increased at more than twice the rate of population growth over the last century. We use a lot of water: 1kg of chocolate takes 24,000 litres, 1kg of beef 15,000 litres, one cotton t-shirt 3,000 litres, and a hamburger 2,400 litres. The demand for water for manufacturing is expected to increase by 400 per cent by 2050. In the UK, river abstraction now accounts for 61% of all water taken from the environment, up from under 40% at the turn of the century. [131] Groundwater use has also risen sharply, up 53% since 2018. These projected water shortages do not factor in climate change and potential increased droughts, fires, floods and changing temperatures. Even places that rely on abundant water, like Tasmania, Australia, rely on rainfall to power its hydroelectricity, but rainfall has been falling for the past 70 years due to climate change. Abundance becomes shortages, and Tasmania now has to import electricity from the mainland and is now a net importer for 11 months of the year.

It is not just quantity but quality. Pollution from industry, agriculture and wastewater contaminates freshwater resources. Of the 86% of inland water bodies that did not meet targets in England, 36% failed due to sewage and wastewater. Yet, the UK is a stable Western democracy. It's even worse, particularly in sub-Saharan Africa and across the Indian subcontinent. The deterioration in quality makes existing scarce water sources unusable for drinking without extensive and often unavailable treatment. Progress has been made in providing access to safely managed drinking water but the rate of progress is insufficient to meet the UN Sustainable Development Goal 6 (universal access) by 2030.

Agriculture accounts for 70% of global freshwater withdrawals. By 2030, global water demand is projected to exceed supply by 40%, driven mainly by agricultural needs. Water stress will increase. While demand will exceed supply by 2030 it gets worse. By 2040 the estimate is that over 40% of all irrigated agriculture will suffer extreme water stress. Agriculture creates other issues, particularly water quality degradation through runoff of fertilisers, pesticides, and animal waste, leading to eutrophication and contamination of water bodies. Pollution creates two problems—the usability of water for other purposes and the long-term sustainability of water sources. Outdated irrigation systems and water-intensive crops contribute to inefficient water use in agriculture, exacerbating shortages.

We are already seeing water equity issues. These water deficits become flashpoints for conflict as water security diminishes. A fair use of water in the Colorado River or the Mekong River exacerbates conflicts. Reduced water supplies in arid and semi-arid regions are projected to displace between 24 million and 700 million people by 2030, increasing pressure on water systems in receiving areas and further fuelling social tensions. Not only do water shortages fuel social tensions, but a lack of water also decreases food security. Food price spikes and instability further contribute to social unrest. Economic costs of water scarcity will impact GDP growth in countries that can ill afford further inequality. Political instability will increase in countries that are already unstable.

10.11 Earth's Resources

The situation is bleak. How do we maintain optimism when we see destruction of habitat, pollution that has gone un-enforced, and species extinction? Are these transformations in renewable energy, transport, or labour going to exacerbate destruction? It is pretty clear that we cannot continue along the same path and keep doing business as usual. Despite the bleakness, we can see optimism. Conservation is a solution for a more regenerative future for people and wildlife. Some argue there is no room for conservation and that we need all the land we have for feeding people. These discussions are usually accompanied by claims that conservation, especially tree planting, is displacing farming. Paradoxically, this is often voiced by the very farmers looking to sell their land. They argue that conventional farming is no longer economically viable and that the next generation isn't interested in pursuing it. Yet still, they blame conservation.

Yet, as we have seen, even in the most populous of countries, urban areas are less than 10% of the land. Fossil fuel mining and its associated waste occupy less than 0.1% of the land area. Likewise, replacing that fossil-fuel energy with 100% renewable electricity would occupy only about 0.18% of the land area. If 50% of existing urban roofs were covered with solar panels, the solar would generate enough electricity to completely electrify the world, so we wouldn't even need new land. Renewable energy systems reduces mining by over 545 times. In Part 1, we also explored that within 20 years, most of the minerals needed for the clean energy disruption will be available from recycled products, which would reduce the minerals we need by up to 90%.

It comes back to the reality that low-intensity grazing in places such as Australia could be easily converted back into protected reserves. Any food production losses from removing those animals on sparse landscapes could be more than offset by moving to more intensive agriculture. What we should be getting angry about is that land clearing has devastated biodiversity and biomass and, in most countries, concentrated that wealth in a small number of landholders and corporations.

We plunder the seas and have crashed marine biomass by 60%, yet continue to pollute the oceans with plastic. As will explore in the health section that 60 kg of plastic we use annually has to go somewhere. We all have over a credit card's worth of plastic in our bodies and eat it each week, yet we recycle only 5% of the plastic produced. Our anger is reasonable when we consider how plastic manufacturers and plastic users have controlled politicians and fought tooth and nail to avoid bottle laws or to impose a cost on virgin plastic to make plastic recycling viable.

There is no technology needed to make these changes. It is only political will and a few simple economic levers that can stop polluters from treating the planet's oceans and land as an open sewer. Chicken and dairy farms use rivers as dumping grounds for nitrates and sewage. We must withdraw subsidies to corporate graziers to ensure that farming marginal land is no longer profitable, so that the land can recover, be reforested, capture CO_2, and serve as a carbon sink.

Likewise we could stop the 11 million tonnes per minute (7 trillion a year) of direct and indirect pollution by fossil fuel companies from treating the atmosphere and groundwater as a free sewer spewing out gigatonnes of emissions. The world has shied away from a carbon price, but a very simple action is to shift existing subsidies (7.1% of GDP) toward investing in zero-emission systems. That would put the world on track to reach zero emissions by 2035.

Using additional land for housing is not the issue. We will get back vast areas of carparks when we change from individually owned cars to robotaxis or shared public transport. There is more car parking area in Los Angeles than there is land area for the whole of San Francisco. In most cities, over 30% of the land is for cars. That area should be replanted with trees to create more livable cities.

10.12 Conclusion of Agriculture

Agriculture is the largest hidden global emitter of pollutants, including CO_2. It emits as much CO_2 as transportation, but we don't see media reports, and the lobbyists are fully embedded in governments in all countries to keep it that way. To make any headway in reducing carbon emissions, the only way will be to stop most ethanol production, switch from red to white meat, stop plundering the ocean, and understand that disrupting food systems is the only way we can stop clearing land for agriculture or further plundering the oceans. Otherwise, most farming actions are merely tinkering at the edges. We can reasonably forecast that changes in food production will drive changes to the agriculture sector, including:

- Electrification of agriculture is well overdue. Solar, batteries and heat pumps will displace fossil fuels for heating, drying and other farm uses.
- Precision agriculture will displace much of the fossil fuels used. It increases yields and decreases crop inputs.
- Disruption of food (next chapter) will reduce crops to feed animals.
- Nitrogenous fertilisers will be reduced. In Part 3, in the hydrogen section, we explore technology from Omics and others that could reduce nitrogenous fertiliser use by 30%. There is much talk and plans to use green hydrogen to displace grey H_2 for the manufacturing of fertilisers. But if plants can synthesise their own nitrogen, we could see a marked decrease in the use of fertilisers, and its unlikely progress will stall.
- Plant bio-stimulants are effective, but have never been adopted as industrial fertilisers are so cheap. Before you think that organic food production will overtake agriculture, note that the fossil fuel industry and big Ag have been working to flood the industry with synthetic fertilisers for decades.
- Electrification of transport spells the end of ethanol for transport. An area the size of Nebraska is currently growing corn for ethanol and will no longer be needed.

Food disruption will drive disruption in agriculture. Remove the animals, and the areas needed for agriculture could be reduced by 80%.

Conclusion of Agriculture:

- Agriculture is the biggest CO_2 emitter.
- Land use for agriculture is over 20% of land, whereas urban land is less than 10% and often as low as 2%.
- Low-intensity grazing is the biggest land user and a global polluter.
- Many industries treat the oceans and land as a sewage disposal mechanism.
- Technology (Electric transport) and Precision Fermentation (Food) are two technologies that will radically reduce farmland by 80%.
- Freed-up farmland is an opportunity, not a loss – for reforestation, biodiversity, and carbon drawdown.

11 AI Impacts for Food

AI Impact on Food:

- From Real Meat to Lab Meat. Like it or not!
- Understanding huge corporations in the West versus micro farm holdings in the global South.
- Will precision fermentation happen, and by when?
- What are the futures for Small Farms versus Big Farms?
- What role is AI playing in this shift?
- Who controls the Western food supply – and what happens when AI and fermentation disrupt their business models?

Food is undergoing the biggest disruption since man started farming seeds, and domesticating animals. While dogs were domesticated over 10,000 years ago, horses were only domesticated into beasts of burden about 6,000 BC [132]. We have seen enormous industrialisation of agriculture and food in this past century, but this is not the disruption. There has been a radical divergence in food production.

- Food is corporatised and industrialised, and just 12 companies supply most food to Western societies. Tracey may think she has a choice, but clearly, she only has the choice of a brand from a few suppliers.
- In contrast, in the global South, there are over 550 million farms under 2ha in size, and these provide 80% of global food, although Ada will purchase most of her food from a few local village farmers.

Concentrating food supply is not the change we see. Rather, the change is from growing crops to feed animals to feeding humans directly. We will explore the technology, the funding, and the predicted changes and consequences of this food revolution. Disruption today is not the plant-based alternatives that look like meat; it's cell-based cultivated meat. While today's plant-based meat is a huge improvement from plant-based food of 30 years ago, the development of meat from animal cells, which bypasses the animal altogether, is the basis for this disruption.

11.1.1 Modern Food

The food market is enormous due to the range of foods and food products modern society has created. Few comprehend that less than 120 years ago the human diet was mostly grain products. Meat was only for the rich and the poachers, and even then, not too much. While cabbages have existed since Roman times, other brassica products, including broccoli, cauliflower, and Brussels sprouts, are 20th-century innovations. Potatoes also arrived in Europe in the 16th century, and carrots as root crops are as recent as the 17th century. Most food in human civilisation was grain-based starch – grain from the Middle East and rice in Asia.

The state of alternative proteins in the food industry is highlighted by the Good Food Institute. [133] Precision fermentation (PF) transition is underway, even if you are unaware of it. In 2023, two companies sold cultivated meat in the USA for the first time, with regulatory approval from both the FDA and USDA, and companies worldwide stepped up development. Companies and countries have doubled their investment. Policies support this transition. If there is one message from this book, it is that PF is here; it has already passed regulatory scrutiny, and as costs continue to fall, it will be in many of your food products within a few years. We will examine the current state of food today, then explore PF technology, its rationale, the market segments likely to be first cab off the rank, and the implications for food and agriculture.

11.1.2 Food Not Poison

Food can be medicine but can also be poison, says Prof Robert Lustig in The *Hacking of the American Mind.* [134]. He points out that in the world, 5% of the global population is clinically depressed, but in the USA, over 22% . Maybe that is why they face some societal challenges. He points to the rise of dopamine from what we eat and explains that fructose has hijacked food. What is emphasised in his first book, *Fat Chance*, is that the dose of sugar (or alcohol) consumed is important. Most dieticians say the maximum daily sugar intake should be under 12g. However, the WHO says 50g, or 10 teaspoons (or 10%) per day, is the maximum. Many Americans consume over 100g per day. The explanation is clear: sugar is spiked or hidden in 73% of all food in America. Over one-third of Americans cannot cook, so they eat food products. Over 45% of adult Americans have chronic fat in the liver, and in children, 25%. Sugar gets turned into liver fat. Lustig suggests Michael Pollan's "*Eat food, not too much. Mostly Food.*" He says it needs to be simplified to 3 words. "Eat real food." The "too much" will take care of itself. Eat plants that come out of the ground. Eat animals that eat those plants. If the food has a label, consider it a warning label because real food does not need a label. More concerning is his warning that during the past 25 years, baby birthweight has increased 200g. Society now has neonatal obesity, and these kids did not get obese from gluttony and sloth. Those overweight babies grow into overweight adults.

We also need to feed the 30 or so trillion bacteria (microbiome) in our gut. What feeds them and creates diversity is fibre. Our ancestors ate 50-100g of fibre per day. While the USDA recommends 25g per day, most Americans and people in the UK consume less than 12g per day, primarily due to their consumption of ultra-processed foods. Fibre cannot be frozen, so processing turns a fibrous food, such as an orange, into a commodity (orange juice). The problem is that one can drink six oranges' worth of juice, but most would struggle to eat six whole oranges at once. The result is that sugar intake goes up; fibre intake goes down.

In this discussion of new food, we must be mindful not to swap one existing problem for another. In my 2014 book *The Takeout Diet* [135] the top 3 takeout actions are to take out fructose from your diet, take out refined carbohydrates, and if a goal is to lose weight, adopt a weight loss program that takes out some food using intermittent fasting. Obesity is not the only problem. Environmental obesogens are widespread. Humans have developed over 100,000 man-made chemicals, including Per- and polyfluoroalkyl substances (PFAS) known to cause obesity and, more recently, cancers. These groups are large, complex synthetic chemicals that have been used in consumer products worldwide since the 1950s. The ingredients in various everyday products include PFAS, phthalates, tributyltin, BPA, plasticisers, parabens in cosmetics, vinyl flooring, and flame retardants in clothes and bedding, all of which are endocrine disruptors. We know that some are carcinogenic and have been banned but there are thousands still in use. The subtle effects on humans' metabolic and endocrine processes are unknown. So, as we look at the evolving food revolution, what will the impacts? We need to move from processed food or nutrients to food. Lustig says there are 3 principles in any diet.

- Protect the Liver
- Feed the Gut
- Support the Brain

Lustig says we have a systemic health crisis, with diabetes, metabolic syndrome, dementia, cancer, a mental health crisis, depression, suicide and a planetary crisis of war, environment and climate change. He blames the food industry for producing food products full of compounds, including fructose, that cause liver damage, dopamine-inducing products that do not support the brain, and a lack of fibre to feed the gut. In this food revolution, we don't want to have food products worse than the ones we have now, nor do we want to export the Western diet to the rest of the world.

11.1.3 Food Industry Now Dominated by Corporations

Few realise the current food industry is now concentrated in the hands of just a few multiple food companies for most countries.

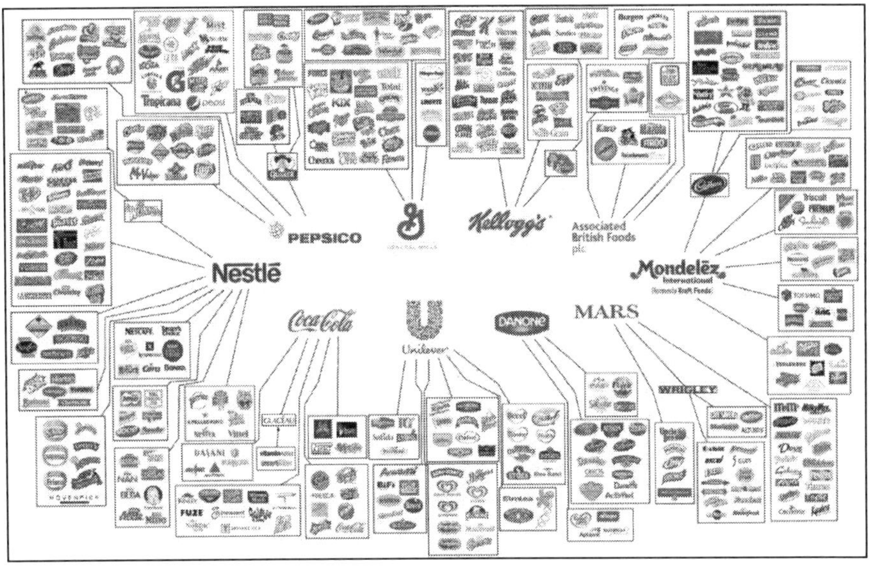

Figure 30 Big brother brands

Figure 30 shows how few companies there are. For Tracey and the readers of this book, just 12 companies own 90% of what we buy. We are familiar with public companies, but the book "*Barons, a story of seven corporate titans, their rise to power, and the consequences for everyone else*", [110] provides insights into the silent private companies. The deregulation of the American food industry has consolidated wealth in the hands of select tycoons and, along the way, hollowed out the nation's rural towns and local businesses. The USA trend is moving to other countries, including Australia. Public companies, e.g., General Mills, own most of the family brands shown in the diagram today. The ten companies include Mondelez, Kraft, Coca-Cola, Nestlé, PepsiCo, P&G, Johnson & Johnson, Mars, Danone, General Mills, Kellogg's, and Unilever, with each company controlling an interest in anywhere between two dozen and almost a hundred other companies, as in the case of Nestlé. Food production is also big business. For example, just 15 fish companies [136] dominate the $257b seafood industry, and just 10 grain producers in the USA dominate grain supply.

Company	Country	Turnover ($b)	Products	Countries	Website
Nestlé	Switzerland	~90+	all	184+	nestle.com
PepsiCo	USA	~79+	all	200+	pepsico.com
The Coca-Cola	USA	~35+	all	200+	coca-colacompany.com
Mondelez Intern	USA	~37+	all	140+	mondelezinternational.com
Danone	France	~28+	all	120+	danone.com
JBS S.A.	Brazil	~46+	Meat	500 plants 24 countries sales 184 countries	jbsfoodsgroup.com
Archer Daniels I	USA	~61+	grain	190+	adm.com
Mars, Incorpora	USA	~37+, private	all	80+	mars.com
Unilever	Netherlands (Br	~50+	all	190+	unilever.com
Tyson Foods	USA	~38+	protein	140+	tysonfoods.com/investors
Cargill	USA	~171+, private	Grain	130+	cargill.com
Walmart	USA	~648+, private	Retail	19+, 10,500 stores	walmart.com
Driscols	USA	~10+, private	Berries	21 grown, 48+ sell	driscolls.com
Marubeni	Japan	~75+, private	Fish	10+	marubeni.com/en
Charoen Pokph	Thailand	~17+, private	Shrimp	10+	cpgroupglobal.com/en/home

Figure 31 Major food companies globally

11.1.4 Small Farms vs Corporations

In direct contrast are the small farms. Globally, in 2014, there were 570 million farms, and in 2021, more than 608 million. [137] These 608m farms were between 70%-80% of the world's farmland and produce the food to 80% of the world's population. You won't find that food produced on small holdings in the supermarkets of the west. In Western cultures, food is concentrated in a few corporations. It's a different world for Ada. According to the FAO [138] farms less than 2ha:

- Five of every six farms in the world (84%).
- 510 million small farms.
- Most family owned.
- 12% of all agricultural land.
- 35% of the world's food.

Smallholders' contributions to food supply vary enormously between countries, with the share as high as 80 per cent in China and in the low single digits for Brazil and Nigeria.

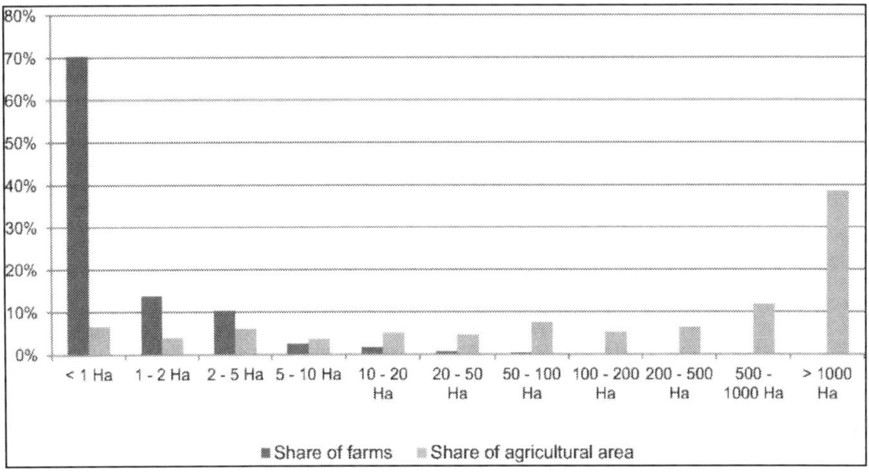

Figure 32 Worldwide distribution of farms and farmland

As shown in Figure 32, the largest one per cent of farms in the world - greater than 50 hectares - operate more than 70 per cent of the world's farmland. [137] Farms over 1,000 hectares are nearly 40 per cent of agricultural land. The higher the average income of the country, the larger the farms and the more concentrated they are.

11.1.5 Food; Not Food Products

If you're looking to avoid contributing to the Big 10 world takeover, try shopping at local farmers' markets and maybe skip out on soda and highly processed foods altogether. The local farmers market are good options, if they exist in your suburb.

11.1.6 We Eat Animals, Lots.

Humanity collectively eats over 360 million tonnes of animals per year. In 2015, global total protein intake was 78.2g/day, ranging from 61.0g in the Middle East and North Africa to 92.5g/day in Asia. Mean national total protein intake was ≥46 g/day in all 185 countries. The recommended protein RDA is 60g per day (i.e. or 20kg per year), and the average American eats over 30% more than that per day and over 85% is from an animal

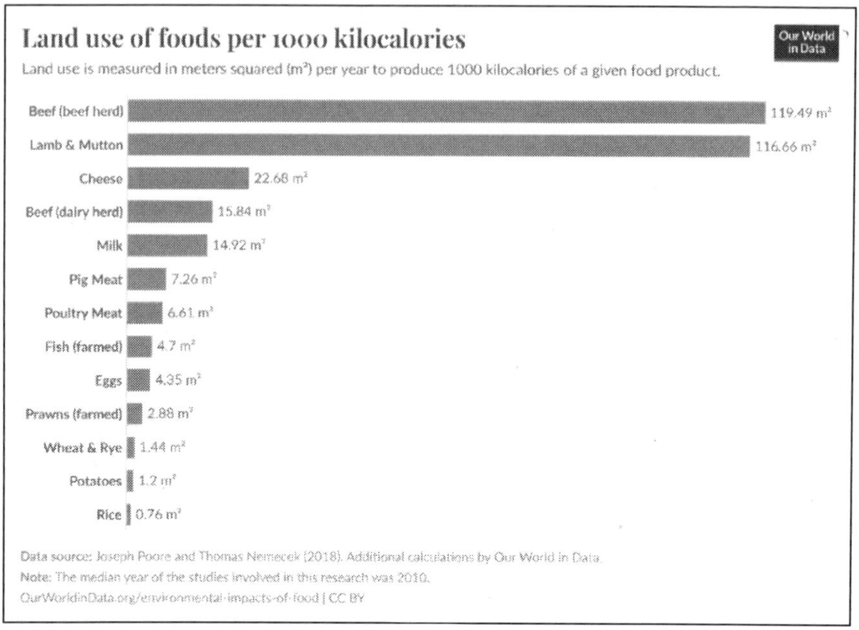

Figure 33 Area required to produce 1,000 kcal of food

Likewise, Australians eat over 90 kg of meat per person pa [139]. The human appetite for protein is also mirrored in the protein content of animal and pet food. Pet food, cats and dogs, in the supermarket has 20% animal content.

11.1.7 Inefficiency of Animal Protein Production

Producing food via animals is extremely inefficient. Figure 33 shows land needed. Beef requires 120 m^2 to produce 1000 kcal. Rice is less than 1 m^2. [140] Farmed prawns use much less land than pigs or poultry. However that is not the whole story. What is probably as important is shown in Figure 34 of the fate of the "phytomass" grown. The inefficiencies expressed as "units of edible output per 100 units of feed input" are lower than the previous feed conversion calculations because they are based on energy flows from plant phytomass to edible animal parts. For example, they factor in feed that is wasted before it reaches livestock, the plant mass used for non-feed purposes such as bedding, the energy that grass and other forage provide for feeding livestock, and the energy needed to support animal production beyond just producing animals. Additionally, many animals do not directly produce food, are culled, die before reaching maturity, or are used in breeding. Wirsenius [140] highlighted 20 years ago the percentage/Units of Edible Output per 100 Units of Feed.

- Poultry – Calories – 11% – Protein 20% (i.e. 9 calories in, 1 calorie out)
- Pigs – Calories – 10% – Protein 15% (i.e. 10 to 15 in, to 1 out)
- Cows/Beef – Calories – 1% – Protein – 4% (i.e. 40 in, 1 out)

When we look at the total phytomass grown in agriculture, we see that only about 8% is used to produce protein.

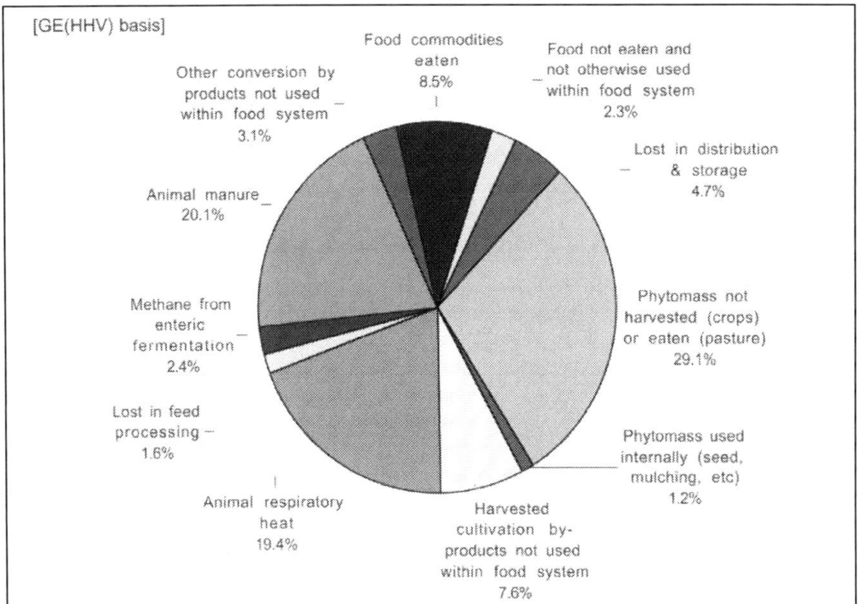

Figure 3.9 Food intake and other fates of appropriated terrestrial phytomass in the food system. World totals. Values are expressed as share of the total amount of appropriated phytomass (221 EJ GE(HHV)/year).

Figure 34 Global phytomass fate of what is grown and used.

Figure 34 shows these huge losses in the food system

- Manure = 20%.
- Not harvested or eaten = 28%.
- Animal respiratory heat = 19%.

If a farmer claims efficiency in agriculture, it is simply false.

11.1.8 Intensive Agriculture and Food Poisoning

We believe that food today is clean and healthy, and that is mostly true, but there are caveats. Food has improved over the centuries and refrigeration has probably been the most substantial improvement. Prior to refrigeration, drying, canning, and preserving provided storage. Food produced in today's industrial environment is generally safe. Risks are downplayed. However, food does contain harmful bacteria, viruses, parasites and chemical substances. They are usually kept under control, but when accidents and mistakes happen they can cause large-scale food contamination events such as:

- An estimated 600 million – almost 1 in 10 people in the world – fall ill after eating contaminated food.
- Foodborne illness, or food poisoning, affects about one in six Americans, or 128,000 hospitalisations and 3,000 deaths annually in USA – CDC. [141]
- Globally 420,000 deaths per year.
- Causes more than 200 diseases from diarrhoea to cancers. [142].
- Creates a vicious cycle of disease and malnutrition, particularly affecting infants, young children, elderly and the sick.
- Loss of 33 million healthy life years (DALYs).
- US$ 110 billion loss each year in productivity.
- *Salmonella* generally come from eggs and poultry.
- *Campylobacter* comes from raw milk poultry and water.
- *Escherichia coli* from unpasteurised milk, undercooked meat and contaminated fresh fruits and vegetables.
- Viruses such as *Norovirus* and Hepatitis A can both be transmitted via food.

Tracey has better practices in food and follows 2-hour/4-hour rule (2 hours out, 4 hours eaten or discarded) but those rules are unfortunately not followed in many parts of the world. Ada and her children suffer frequently from food contamination and lack access to potable water. Her children are at high risk of death from disease and malnutrition. Any food disruption should reduce risks.

11.2 Precision Fermentation

Precision fermentation (PF) relies on multiple technologies, termed "Precision Biology" which is the coming together of modern information technologies like artificial intelligence (AI), machine learning, and the cloud, together with modern biotechnologies such as genetic engineering, synthetic biology, metabolic engineering, systems biology, bioinformatics, and computational biology. Technologies enabling food disruption include AI, genetics and energy. Terms used include:

- Precision Fermentation (PF), which, in addition to traditional fermentation (e.g., beer, bread), includes the cultivation of a microbial species for either whole-cell biomass (biomass fermentation) or for the production of functional food ingredients by recombinant protein technology (PF).
- Cultivated meat, or tissue-engineered cellular agriculture, which involves growing live cells from an animal and then directing these cells into a desired product (e.g., sausage, mince patty, meat).

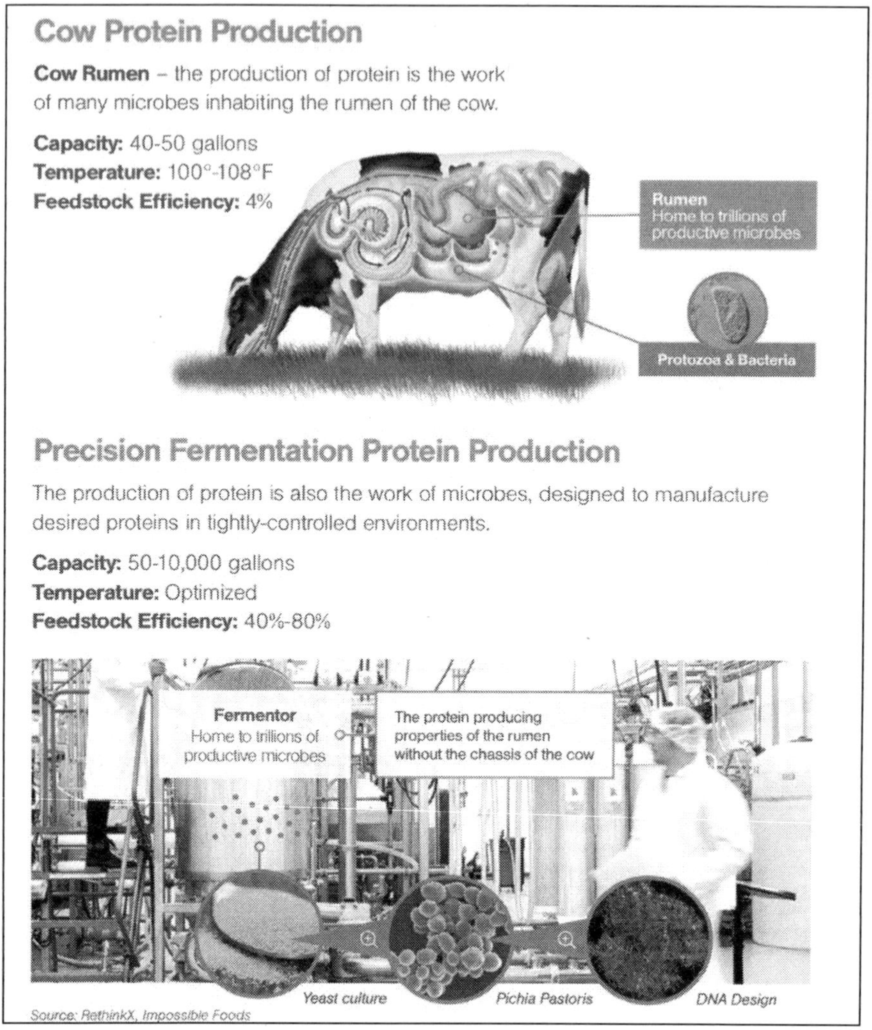

Figure 35 Cow versus precision fermentation to produce protein

Figure 35 from RethinkX compares a cow with a PF for milk protein (casein) production. Conversion efficiency has increased from ten to twenty times from 4% bovine efficiencyPF will be a range of techniques enabling us to produce almost any complex organic molecule. The fermentation of proteins in a digester (like a craft beer tank) with a sugar (or starch) feedstock and precision-biology bacteria or yeasts. The result could be proteins, fats, sweeteners, or cell-based meat.

RethinkX above uses the example of PF for cow protein, and shows PF costs have decreased exponentially. They use insulin as a case study. [143] When insulin was first produced, a kilogram of insulin was extracted from over 50,000 pancreases of cows and pigs. In 1982, Genentech produced insulin using modified yeast. In 1995, PF became competitive for medicines. In 2010, PF became competitive in cosmetics at $10,000 per kg and in materials at $1,000 in 2015. At $10 in 2024, PF becomes competitive with animal protein. By 2030, PF will be under $5, and by 2035, potentially under $ 1 kg.

Company Name	Country	Type of Food	URL
Aleph Farms	Rehovot, Israel	Cultivated Steak	aleph-farms.com
BlueNalu	California, USA	Cultured tuna toro	bluenalu.com
Change Foods	USA (was Aus)	Animal-free dairy (cheese and other products)	changefoods.com
Finless Foods	California, USA	Cultured tuna	finlessfoods.com
Formo	Germany	Animal-free dairy (cheese)	formo.bio
Geltor	USA	Collagen and other high-value proteins	geltor.com
Imagindairy	Haifa, Israel	Animal-free milk proteins	imagindairy.com
Infinite Roots	Hamburg, Germany	Mycelium to soy like protein	infiniteroots.com
Meatable	Delft, Netherlands	Cultivated pork	meatable.com
Meati	Colorado, USA	fungi meat protein	meati.com
Melt & Marble	Sweden	Designer fats for meat alternatives	meltandmarble.com
Mosa Meat	Maastricht, Netherlands	Cultured burger	mosameat.com
Motif FoodWorks	USA	Animal-free proteins, fats, flavors	madewithmotif.com
Perfect Day	California, USA	Microflora-based whey proteins	perfectday.com
Remilk	Rehovot, Israel	Cultured milk and dairy products	remilk.com
The Every Company	California, USA	Cultured egg white protein	every.com
TurtleTree	Singapore/USA	Sustainable Lactoferrin	turtletree.com
Upside Foods	California, USA	Cultivated beef meatball	upsidefoods.com
Vivici	Netherlands	Animal-free dairy proteins (β-lactoglobulin)	vivici.com
Wildtype	California, USA	Cultivated salmon	wildtypefoods.com

Figure 36 Sample of Precision Fermentation companies

In Figure 36 we see the scale of businesses investing in PF.[144]

- 165 fermentation companies focused exclusively or predominantly on alternative proteins by the end of 2024.$651 million raised in total private funding in 2024.
- Over $4.8 billion in all-time investment in PF.
- 17 fermentation facilities were opened or announced in 2024.

The four largest fermentation deals in 2024 were Perfect Day's $90 million Series E, Formo's $61 million Series B, Infinite Roots' $58 million Series B and Meati's $100 million Series C, which went into administration in May 2025. The investment climate has cooled in 2026 due to the focus on AI from investors.

Some of the products, such as Beyond Meat, launched in 2009 with burgers and, after launch, saw their valuations soar to $220, but now they are at $3 per share. It is likely headed for bankruptcy by 2027 due to $1 billion in convertible bonds it owes, but its stock has retained some value. Most of the newer companies are providing their products to existing food producers with established marketing channels. Of the 170 companies around the world that will go out of business or consolidate, it is anyone's guess.

11.2.1 Rationale for PF

As Bruce Friedrich, founder and CEO of the Good Food Institute, said in an interview [145] that the central brainstorm is:

- Meat is made up of lipids, amino acids, minerals and water. That is everything that constitutes meat.
- Plants also have lipids, amino acids, minerals and water.
- Plant protein is much more efficient than animal protein; nine calories into a chicken to get one calorie back out, 40 calories into a cow to get one calorie out.
- Therefore, hire scientists, chemical engineers, biotech specialists, and mechanical engineers. We should be able to figure out how to replicate the precise taste, texture and everything else of meat using plants so that it is indistinguishable.
- Because it is much more efficient, we should be able to get it to a place where it costs less, is more nutritious, has lower fat, lower saturated fat, more complex carbohydrates, some fibre, and no cholesterol.
- PF technology is similar for cultivated meat as it is with plants.
    - Plants - take a seed from a plant or a cutting from a plant, and grow it into an entire plant, or with tissue culture, take a cell and multiply it up to a plant.
    - Meat - take cells from a chicken or a pig or a fish or any animal, you can bathe the cells in nutrients and grow it into actual animal meat.

The difficulty with animal protein compared with plant tissue is the complexity of the connective tissue. Plants have phloem and xylem, but animals have over 2,000 cell types. Animal tissue has a more complex structure.

11.2.2 Advantages of PF

The advantages of PF include:

- More efficient.
- No killing of animals.
- Less land, and therefore less fertiliser, pesticides, and fossil fuels for farming.
- Healthier product.
- No antimicrobial drugs.
- No contribution to pandemic risk.
- No bacterial contamination.
- If it's seafood, no dioxins or mercury.
- No methane. Animals contribute more methane than coal and gas.

11.2.3 Efficiency Of PF

In contrast to the inefficiencies in our current farming, converting simple carbohydrates to protein in a vat is 10 to 50 times more efficient. Scaling the industry is challenging. While we have established tissue culture for human therapeutics, it is expensive and not scalable. Food production will need to be orders of magnitude bigger. Currently, PF uses pharma-grade feed to grow human tissue, and the change is to switch to food-grade feed to grow animal tissue. It needs to go from very small volumes in a petri dish or a small bioreactor to something at the 50,000-, 200,000-, or even a million-litre scale.

Immortalisation of cells is an issue that has been addressed. If you immortalise the cells, you will need techniques that comply with genetic engineering regulations. However, if you have a small herd of donor animals, scientists can take a small biopsy, maintain genetic diversity, and avoid regulatory complications.

11.3 Disruption of Food

RethinkX projects [143] protein will be cost-competitive with animals by 2030, and 10 times cheaper by 2035. They outline how novel technologies like PF and cellular agriculture would disrupt and transform our food systems, creating decentralised, equitable and environmentally friendly modes of food production, and better-tasting, healthier food! They call this model of food production Food-as-Software (FaaS).

Measure	2030	2035
Price of casein protein	1/5th Cost	1/10th cost
Cost of food product	1/2th cost	1/3rd cost
Household food costs		$1,200 annual cost reductions
Dairy Products demand	70% reduction	90% reduction
Dairy Herd	50% reduction	90% reduction
Beef Industry	10% reduction	90% reduction
Land use	30% less land farmed	60% less land farmed
Greenhouse emissions	45% reduction	90% of demand reduced
Water Use	50% reduction	75% reduction
Waste		90% reduction
Farmland value	40% fall	80% fall

Figure 37 Precision Fermentation Impacts from RethinkX

Companies launch products at a higher price point than their animal counterparts to cover the high cost of creating precision fermentation food that delivers on taste and nutritional value but it is expected over time that PF foods will be cheaper than existing food products.

11.3.1 Disruption Factors

PF will disrupt various food sectors at different times and varying degrees depending on:

- Ability to develop appropriate PF technology.
- Product substitution.
- Economics of production.
- Market approach. Will it be business-to-consumer, business-to-business? Can PF products substitute directly into the food manufacturing process?
- Politics and pushback from the existing industry. Opposition will be fierce. Livelihoods at stake. National interests and geopolitics will become major concerns. States such as Florida have already enacted

legislation to ban "fake meat", so concerned are farmers about the long term viability of the meat industry.

11.3.2 Widespread Changes

Don't make the mistake of thinking this is decades away. PF products are on the market already, and you may well have consumed PF products today. Those in PF disruption space cannot overstate the potential impacts of these changes - not just for farmers but also consumers and animals.

- All livestock markets, including chicken, pig, and fish, will follow a similar trajectory.
- Half of the 1.2 million jobs in U.S. beef and dairy production and their associated industries will be lost by 2030, climbing towards 90% by 2035.
- RethinkX states the USA precision fermentation industry will create at least 700,000 jobs by 2030 and up to one million jobs by 2035. This is probably wrong. Humanoid robots will be cheaper than human labour.
- Oil demand in agriculture will halve.
- About 95% less land is required for protein production from PF than from cattle. Those 200 million ha (485 million acres) are 13 times the size of Iowa, 6 times the size of Germany, or 1/5th of China.
- Crop farming will be reduced by 50% from 120 million tonnes to 80 mt in 2030.
- US beef and dairy cattle consume about 50% of the crops produced for US livestock – 70% of the hay, 45% of the corn, and 17% of the soy.
- If forested, it would offset all carbon offsets for other areas of the economy.

Separate to the disruption from PF will be the disruption in farming due to the replacement of biofuels with batteries for vehicles as discussed in the biofuels chapter in Part 3, and the previous section on agriculture.

11.4 Dairy Industry

RethinkX suggests the dairy industry is ripe for disruption, as protein accounts for only 3.3% of milk. PF flips the industry on its head. With cows, we start with the complex of the whole cow, then break the cow down into the whey, casein and milk fat. That lifecycle takes 2 to 3 years from insemination, calving, maturation and harvesting. With PF, you start with the single molecules and build up the structurally complex products. Time? Weeks. Feedstock efficiency is 40% to 80% versus 3% from feed to product. E.g. beef, you require 40 calories for one calorie of beef or protein along with vast quantities of water, fertiliser and land, and then you have to dispose of the manure and waste products.

By 2030, the number of cows in the U.S. will have fallen by 50% and the cattle farming industry will be all but bankrupt. All other livestock industries will suffer a similar fate, while the knock-on effects for crop farmers and businesses throughout the value chain will be severe.

Tony Seba, RethinkX https://www.rethinkx.com/food-and-agriculture

11.4.1 4 Ways This Disruption Process Will Play Out

- Substitute ingredients. The first wave of disruption will be to change casein from cows to casein from PF/
- Substitute the food end products.
- Fortification of foods.
- Change in form factors.

Substitution of ingredients is a business-to-business. About 40% of casein is used in food products, so the consumer will be unaware of the change. E.g. replacement of whey protein in sports drinks or baby formula, or of gelatine, a common ingredient used as a thickener.

11.4.2 Ramifications for Cattle Industry Value Chain

No part of the livestock industry is spared. Changes will be profound.

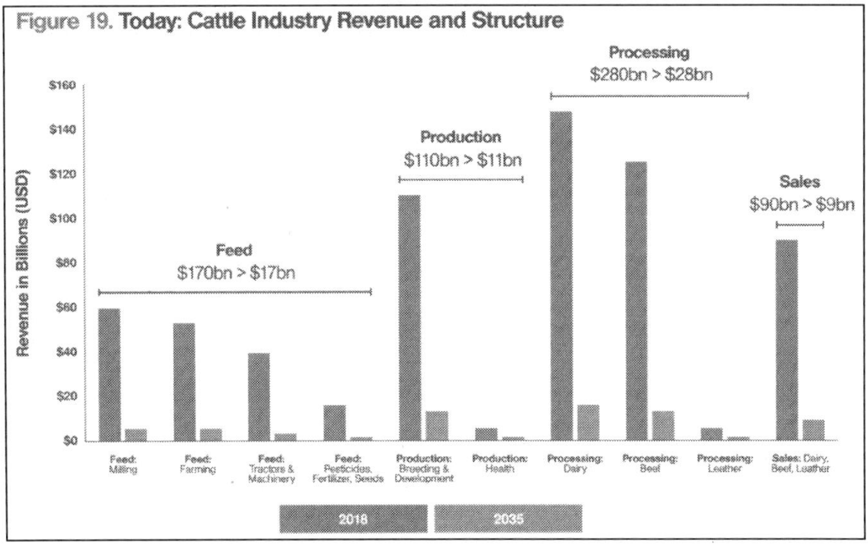

Figure 38 RethinkX – Cattle industry revenue 2018 to 2035

What will the land used for dairy production be used for? Even without active reforestation, the passive reforestation of land through natural recovery will capture and store carbon equivalent to up to 20% of today's global emissions. A number of land use options are available such as:

- Urban and suburban development and conservation. No need to protect farmland so allow small hobby farms.
- Used to restore wildlife habitat.
- Safeguard biodiversity.
- Improve water quality.
- Combat climate change through reforestation and carbon sequestration.

11.5 Meat

An often-insurmountable price barrier is keeping many people from buying plant-based alternatives to beef, pork, and chicken. Plant based meats cost an average of 77% more than conventional meat, and are more processed as they are derived from soy and pea protein. [146] Those who consume existing plant-based meat products are young and non-white consumers. Only 3% of households buy plant-based meat products, which are generally about 20%

more expensive. A challenge that alternative proteins face is overcome government subsidies for meat, in the form of family benefit welfare.

11.5.1 Ground Beef Precision Fermentation

Over 45% of beef consumed in the USA (12.6 million tonnes) is ground beef (mince), and over 60% of prepared meals eaten at home contain ground beef. So, what will change? It is worth revisiting those three views about innovation and change: technology, economics, and belief systems.

Ground beef is the easiest to duplicate using cell beef. Glucose is less than $0.50c/kg. Ground beef is about $5/kg at the farm gate, and then you have processing costs on top. The consumer will pay about $10/kg for ground beef. Steak is more than $20/kg. When the cost to produce cell beef comes under $5 and approaches the cost of glucose by 2030, why would one of those 12 major corporations release a ground beef option?

- No antibiotics.
- No meat processing plants.
- Consumers cannot tell what product is in that store-bought lasagne.
- No price volatility for beef or feedlot costs.
- As costs come down to $1/kg by 2030 expect the industry to be in crisis management. Note that Magic Valley has FSANZ approval for mince. [147]

As well as the changes in dairy and beef production, products such as skin for leather and materials will also be disrupted by PF in collagen and other products.

RethinkX predicts a massive change from today to 2030. They predict that by 2030, just five years away, the ground beef market will decline by 70%, and dairy cow numbers by 90%. Those are big numbers, and it is probably over-optimistic because of the investment environment and political backlash.The current ramp-up is constrained because companies have shifted their focus to producing weight-loss proteins to meet demand for Ozempic and similar products. Additionally, the beef industry has invested in political opposition. Driven by the vast quantities of ground beef consumed, cost reductions, and political argie-bargie, including tariffs by the USA, it will simply be a matter of time before this plays out.

Segment	Reduction by 2030
Beef Cattle	
Beef - Ground beef	70%
Beef Tissue Beef (steak etc)	30%
Total Beef	50%
Beef Herd	28.3 million
Dairy Cows	
Ingredients	90%
End Products	85%
Leather and Materials	90%
Total Dairy	90%
Dairy Herd	9.3 million

Figure 39 Impact of precision fermentation foods in the USA to 2030

11.5.2 Niche Meat

Vow foods [148] have lab-grown meat, approved for sale in multiple countries. They grow quail livers and turn them into foie gras – normally made from chicken liver. While the product is approved for sale in Australia. Italy and some states in USA have banned it.

11.6 Chicken

There is both a meat industry and an egg industry. Efficiencies and disruptions will impact both industries. Chicken meat production is more efficient than dairy or cattle production, but it still takes over 9 calories to produce 1 calorie of chicken meat. While an egg is about 80 calories it takes a range of other inputs to get it to market.

- Takes about 100g of feed per day from the start of egg production at 32 weeks and continues for about a year.
- Water.
- Antibiotics.
- Disposal of chicken manure.

- Egg distribution.
- Continuous turnover of layers – need to bring in new laying hens and dispose of the older less efficient hens. There is a whole industry to breed new hens and raise them before they become productive laying hens.

11.6.1 Egg Precision Fermentation

Development of PF eggs is not going to be a consumer change. People like their fried, scrambled or eggs benedict, so this won't change for the billions of eggs in India (the worlds largest consumer of eggs). However, it will change in food manufacturing. About 60% of the eggs are used in baking and pie filling. At the consumer end, customers won't see any difference when bakeries change from eggs to PF egg proteins. As RethinkX points out, each part of the egg serves a different purpose, such as gelling, foaming (egg white), and emulsifying (egg yolk). It is the egg white that contains the functionally important proteins – ovalbumin (54%), ovotransferrin (12%), ovomucoid (11%), ovomucin (3.5%), and lysozyme (3.5%). [149] The price of egg protein from hens is about $7 to $ 10 per kg. RethinkX believes that a comparable price of PF has arrived, and companies such as Agrionomics with Onego Bio provide a bioidentical protein for conventional egg white, an easy switch for food formulators. It is a $270 billion global egg market, growing to a $387 billion by 2028. All it takes is another outbreak of bird flu (avian flu), and an egg shortage, and the industry will rapidly change.

11.6.2 Chicken Meat Precision Fermentation

The chicken industry likewise is undergoing rapid change. [150]In Figure 40 we see USA produces over 9 billion chickens for about 20 million tonnes. It exports about 17% of production. China has rapidly increased production since 1990. The rise of meat production in India and Indonesia demonstrates how, as countries increase GDP, their protein consumption soars. The USA numbers are still rising as they change from beef to chicken and munch down on 42kg per person per year. A large percentage of chicken meat is "chicken tenders", such as Chicken "McNuggets." As an aside, only about 50% of tenders is chicken! The rest is batter, seasoning and oil.

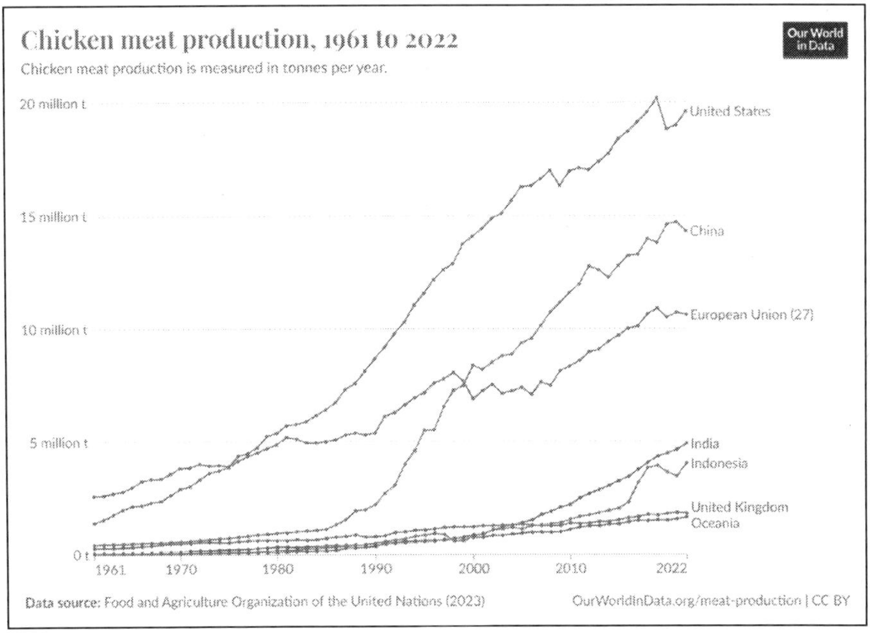

Figure 40 Poultry production 1961 to 2022 Our World in Data

In Figure 40, we see the continued rise of chicken meat production. Even in the USA, which is a "mature" market, chick production continues to rise. Chickens can take comfort that each one will provide enough for about 12 "Happy Meals." And 12 happy parents! About 42% of chicken is sold through food service outlets. (if you recall, over 70% of all food consumed by Americans is a food product!). With over 25,000 farmers in the USA and 1.2 million employed in the chicken meat industry, the advent of cell-cultured chicken for those tenders, or chicken mince in Asian dumplings, will wreak havoc on the industry. It won't take much, as fewer than 30 companies in the USA control the industry. All it needs is just one company to adopt a lower-cost model, and others will follow.

The move in the Western world from cage eggs to free-range eggs for bird-cruelty reasons also brings a higher associated cost of hens. Production of egg-laying birds kills all the male day-old chicks, as a laying chicken has different genetics from the meat chicken, and vice versa. Similar to the egg industry, day-old female chicks for meat are also killed on the first day of their life. That's 12 million just in Australia, over 60 million in the USA. An estimated 7 billion globally are minced and find their way into fish meal or fertiliser. The animal protection societies will welcome the reduction in animal deaths, and environmental activists will welcome any reduction in manure and subsequent contamination of waterways.

11.7 Fish Precision Fermentation

Using fish PF has major implications for both protein production and ocean health. Billions of people rely on healthy oceans for jobs and food, and demand for fish protein is only increasing. Projections show growth of 14 per cent by 2030 versus 2020 levels, driven by growing markets in Asia, Europe, Latin America, and Oceania. Since the 1990s, global catches of wild-caught seafood have declined by about 1 per cent per year, primarily due to overfishing, with more than 85 per cent of the world's fisheries pushed to or beyond their limits. We have denuded oceans with over 60% of the biomass gone over the past 100 years. [151] Most projections indicate that biomass will continue to decline with climate change. [152] In most countries, fish catches are down to pitiful levels. Many countries now rely on aquaculture for additional fish protein, but that comes with its own set of environmental challenges.

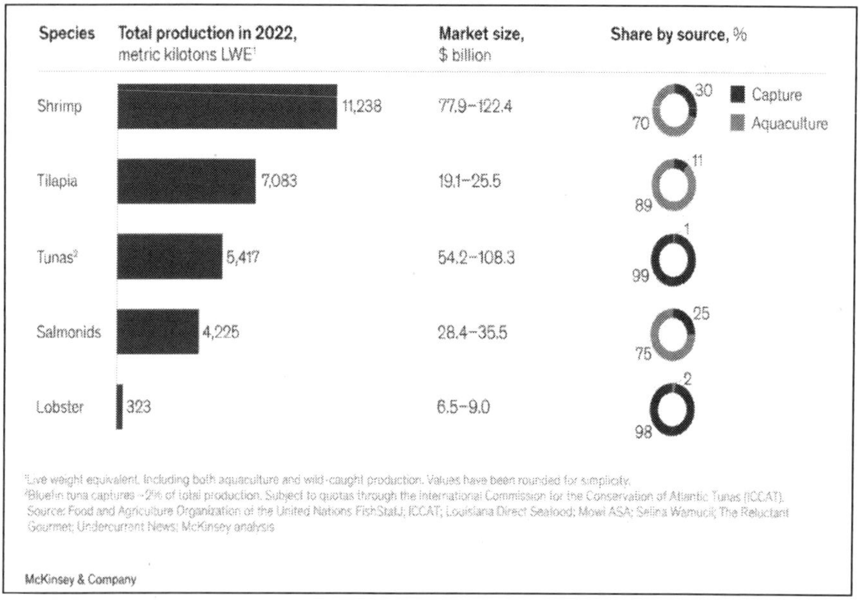

Figure 41 Several popular seafood varieties are attractive for alternative production.

A recent mass die-off of salmon farming event in Australia [153] saw scrambling by environmental groups, industry and government. It's a recipe for disaster as sea temperature rises and farming intensifies. In Figure 41, McKinsey & Company show shrimp at 11.2 million tonnes, tilapia, tuna and salmon are the market leaders. [154]. There are 3 primary production methods.

- Plant-based using soy, seaweed, yeast, legumes, and various vegetable oils and starches
- Fermentation - enabled. Traditional fermentation, biomass fermentation, typically algae, and precision fermentation.
- Cultivated - precision fermentation, but cells grown on biocompatible scaffolds or 3d printed to create fish protein.

As with chicken meat, much of the fish protein is used in foods where its physical characteristics are hidden. Where fish protein needs to be presented, PF plus 3D printing will create the fish fillets. Unlike beef, fish protein is more regular. Most have little clue what fish is used in burgers, fish fingers or Asian dishes. The best they can say it is white and bland.

Israel's Steakholder Foods (formerly MeaTech 3D) has brought us a lab-grown steak [155] and lab-grown Wagyu beef morsels [156]. Now, it has produced what it says are the world's first whole fillets of cultivated fish grouper. [157]. The partnership between Stakeholder Foods and Singapore's Umami Meats created the grouper fillets.[157] The Umami Meats supplied natural grouper cells, which the Stakeholder cultivated and incorporated into a 3D-printable bio-ink. They are bringing that to market. In the USA, there are plant-based fish products from companies such as Aqua Cultured Foods that use algal protein bases.

11.8 Sweeteners PF

Sweeteners made by precision fermentation are already in production. As PF impacts the egg, fats and protein market, sweeteners are probably ahead. Humans are getting bigger and more obese, with consequent lifestyle diseases. Hundreds of books urge restricting sugar intake. If you haven't read one of these, where have you been? Most countries have increasing rates of obesity and medical costs from lifestyle diseases. World Health Organisation estimates that about 2 billion people in the world, including tens of millions of children, suffer from overweight and obesity, more than from malnutrition.

11.8.1 Sugar Production Snapshot

The sweetener industry is a very large industry and the “candy” isle is second only to the pet food isle in supermarkets of being the most profitable isle.

- The global sugar market size 2022 = $195 billion.
- Expanding at CAGR =2.72%, or ~ $230 billion.
- Sugar substitute market (2023) = $17 billion or $28 billion by 2030.
- Sugarcane production 2020 =~ 1.87 billion tonnes.
- Sugarcane 80%, sugar beets 20%.
- Brazil = 40%, India 20%, China 6%.
- Seven countries devote more than 50% of their land to sugarcane cultivation.
- Cultivation area = 26 million ha.
- Average sugarcane yield = 71t/ha.
- Sugar is 10% of the yield, molasses 4%.
- 81% goes to sugar, the balance to biofuel.
- A 1% take from artificial sweeteners would be 260,000 ha.

11.8.2 Why Artificial Sweeteners?

Added sugar is part of the rise of obesity. It is not the only factor, but food companies are looking for sweeteners to add to food products to entice people to consume more. Lower calorie sweeteners, with reduced glycaemic impact, are in demand. Consumers like sugar, and most won’t stop, as they have become addicted. The global sweetener market in food and beverage is a multi-billion-dollar industry and is continuing to increase. PF enables the production of new sugars – taste without the calories. Most are aware of the original aspartame, stevia, and other sweeteners used in food products and many want alternatives.

The quest to find healthy and cheap sweeteners is explored by the BBC [158]. There are multiple patents for food ingredients. [159] Artificial sweeteners have challenges, such as aftertaste and mouth-feel, which can be problematic. Sugar also plays an important part in texture, browning colour, and shelf life – roles that alternatives can fail to fill. Examples of sugar substitutes include:

- **Allulose** is a low-calorie sugar alternative with 70% sweetness of sugar, negligible impact on blood sugar, and found naturally in small quantities in figs and raisins. It has been produced from fructose, which is expensive, but Ambrosia Bio uses precision fermentation to create it from sucrose. The structural difference between D-allulose and D-fructose is located at the paired C2–C3 atoms.
- **Tagatose** is found in dairy products, formed when milk is heated. It is similar in texture and appearance to sucrose but is 92% as sweet but with only 38% of the calories. While it has been around since 1988, lower production costs make it attractive. US-based start-up Bonumose, with the assistance of ASR Group, the world's largest refiner of cane sugar, opened a new plant to produce at lower cost.
- **Erythritol** is found naturally in some foods. It's also made when things like wine, beer, and cheese ferment. Besides its natural form, erythritol has also been a man-made sweetener since 1990. The FDA says erythritol is good for oral health because it slows the growth of one type of bacteria and decreases the acid that bacteria make.
- **Supplant – Sugars from Fibre** New bulking sweeteners from the Supplant Company are a low-calorie (50% lower than sugar), low glycaemic response product which is mildly sweet. Supplant produces it from agricultural waste, including cobs, husks, stems and stalks, using enzymes found in fungi.
- **Silica Anti-Caking** An Israeli start-up, Incredo, embeds sugar crystals with the inert mineral silica (sand). Silica is commonly used in small amounts in food. E.g. an anti-caking agent
- **Sweet Proteins** Six sweet-tasting proteins (thaumatin, pentadin, curculin, mabinlin, brazzein, monellin) come from tropical plants as does the taste-modifying protein miraculin. These proteins are thousands of times sweeter than sucrose.

11.8.3 Growth of Sweetener Businesses

Most days, there is an announcement of company advancement and examples include:

- US-based start-up Oobli produces these sweet proteins (brazzein) by fermenting sugar using genetically modified yeast.

- Abu Dhabi company Novel Foods has signed a $500 million production deal with Abu Dhabi. (Novel Foods). The deal was signed by Novel Foods Group, a subsidiary of the investment conglomerate Gewan Holding; Glatt Group, a German expert in plant engineering; BioEngineering AG, a Swiss company specialised in engineering and design; and CB Consultancy AG, a Swiss company providing engineering solutions for cutting-edge industries.

11.9 Pet food

USA pet owners spend the most on pet food, paying an average of $137 per month, Canada spends $72, and Australian pet owners spend $57. In the USA and UK, 20% of animals killed go into pet food, one of the highest margin products in a supermarket. While one could argue that it is only waste protein from human meat supply that goes into pet food, RethinkX points out the $24 billion pet food market accounts for about a quarter of America's total animal-derived calories. In fact, 160 million pets in the U.S. consume so much meat that, if they were their own country, they would be the 5th largest consumer of meat in the world.

RethinkX said [160] In 2019, in their Food and Agriculture report, the report stated that the pet food market is likely to be the first where cell-based meats are widely used and, because of its size, it will take material profits away from the animal-derived meat industry, thereby accelerating the wider disruption. Examples of companies include:

- Green Queen, BioCraft (formerly called Because Animals) announced cell-based mouse meat reached price parity with premium pet food according to a recent report [161]. It said its cultivated meat now has a sale price of $0.80 to $1 kg. They claim parity with premium meat in pet food and aim to have their products on sale in 2026. They already have chicken cell-based products, but have developed a mouse-based product as a novel product.
- Meatly from Agronomics in the UK uses cultivated chicken as the protein source. They contain no serum, steroids, hormones, growth factors or antibiotics. These products have to go through FDA veterinary medicine processes. Their cost is about $1.30/l

- Pulse in the UK sells dog food from pulse, algae and yeast proteins.
- Cult Food Science has a high protein made without industrial farming.
- Bene Meat Technologies have EU approval for their cultured meat.
- Loop worm from India uses silkworms and black soldier flies for protein.
- Enifer from Finland will create mycoprotein by 2024 as drop in for food producers, and a fish food protein by 2027.

There will be a whole rash of new companies, and with price parity and further reductions will stop animals being killed for other animals. How fast will this ramp up? Most would expect that it will be a minority by 2030 and mostly changed by 2035. If that is not good news, what is?

11.10 Fats and Oils

When talking about PF, we forget humans have hundreds of food products, most which have fats and oil ingredients. We are familiar with olive oil or canola oil but there are dozens of other oils. A start-up in Estonia, Äio, developed a unique process that enables the food industry to reduce its dependence on harmful animal fats, palm oil and other vegetable oils. Using special yeast, Äio converts wood and agricultural by-products, such as sugar from sawdust, into food-grade fats and oils. [162]

The palm oil industry is helpful to see how disruption will play out. Wilmar, Sinar Mas, and Musim Mas have Indonesia sewn up, and Sime Darby Plantation Berhad and Kuala Lumpur Kepong Berhad, Malaysia. Along with Cargill, Bunge, ADM and some tycoons, the palm industry is controlled by less than 20 suppliers and users.

Big brands such as Unilever and Doehler are investing in precision fermentation-derived palm oil in a bid to reduce the environmental impact associated with conventional palm oil, a $70 billion market mostly in Malaysia and Indonesia. [163] Palm oil has become a mainstay in a huge range of products from peanut butter to pizza, soup to soap, thanks to its versatility and low cost. About half of the global biodiesel production is used in the EU. RethinkX explores this example [164] of palm oil in a lotion. The lotion may contain 70% water and 30% functional ingredients, of which 5% is palm-derived. The manufacturer switches that small amount of palm oil and customers are unaware as the label still says palmitic acid. PF will disrupt that 3% of the bottle, but 100% of the market for those incumbent oil producers. Furthermore, drive in an EV to the store to pick up that lotion, and you have killed the demand for palm oil – both diesel and lotion. That has devastating consequences for the growers of palm oil, and the only solace is that most of those are major corporations.

11.11 Animal Welfare

Groups such as PETA and animal rights groups wage campaigns against animal cruelty, especially regarding their conditions during their lives or at the end of their lives. Vegetarians say we should not eat animals. Humans consume more than 1,000 animals in their lifetime, although it may be much more than that. Tracey and her family might wolf down a dozen prawns or two dozen oysters at festive events in one sitting, whereas Ada and other billions like her eat minimal protein! We are all aware of the number of animals killed in industrial agriculture, whether in meat chicken or in egg production, that 50% of all day-old chickens are culled (killed), whether females for meat, or males for egg laying. In dairy farming, huge numbers of bobby calves are slaughtered. Wherever there is industrial farming, animals are killed and often used to feed other animals, sometimes with severe human consequences, such as VJC disease (mad cow disease). Moreover, our desire for animal protein has led to the rise of zoonotic diseases. Zoonotic diseases are infectious diseases that can be transmitted from animals to humans. Many new human diseases originate in animals, and some zoonotic diseases can be severe, even fatal, for humans.

Every day, we kill 900,000 cows, 1.4 million goats, 1.7 million sheep, 3.8 million pigs, 12 million ducks, 202 million chickens, and hundreds of millions fish. [165]

11.12 Pushback

Everyone knows change also brings detractors. From computers, smoking, renewable energy to electric vehicles, some push back. When livelihoods are at stake, farmers and agricultural producers will try to retain agricultural subsidies and fight off lower-cost protein. Agricultural subsidies in the twentieth century were designed initially to stabilise markets, help low-income farmers, and aid rural development. The cost has been tremendous, and the impacts are horrifying from an ecological perspective. From the mid 1980's farm income from support income has dropped in the USA from 22% to about 10%; EU 49% to 19%; Canada 35% to 10% and Australia from 12% to 3%. While that sounds good, in real terms over half a trillion ($540 billion) has been given to farmers every year between 2013 and 2018 in global subsidies. Are subsidies good or bad? It all depends [166] on the study. E.g. China argues that agricultural subsidies boost farmers' willingness to cultivate grain, reduce land abandonment and that the subsidies have increased sown area, output, and grain income. [167] In many countries, subsidies for agricultural inputs, diesel and fertiliser will hinder climate change policies.

11.12.1 Farmer Lobby Groups

We see farmer lobby groups already marshalling against non-farm food products. Bans on food products with names such as "milk". Often, they have lost the battle over product naming. Florida Gov. DeSantis says the ban on lab-grown meat not only protects the state's cattle farmers but also thwarts the "authoritarian goals" of the global elite to make us eat "fake meat". DeSantis was particularly incensed after Bezos invested in lab-grown meats. Florida became the first US state to outlaw the manufacture and distribution of lab-grown meat, but other states may do the same.

Whole supply chains for industrial farming are going to be disrupted. E.g. nitrogenous fertilisers are currently manufactured with fossil fuel. About 87% of anhydrous ammonia is used as fertiliser, with the remaining 13% used in chemical and industrial sectors. The ammonia industry will no longer have that demand. Green hydrogen is touted to decarbonise the ammonia industry, but food disruption will obviate much of that need. Expect to see farmer lobby groups ramp up media. Loss of farming, loss of subsidies, loss of income from farm suppliers are powerful drivers to prevent change in the industry.

RethinkX says that the disruption of the cow will be in place before before technologies can produce the perfect steak, and this is due to both a virtuous and a viscous cycle in the change. The beef and dairy industries operate on extremely thin margins, with high operating and financial leverage, and are propped up by government subsidies. Both are already hanging in the balance and just a small drop in demand will send them spiralling towards bankruptcy. PF will not make much difference to the global 550 million small family farms, but will make a huge impact on the dozen corporate farmers. They will recognise the challenge, try to gain a share of the new market, and slow the rate of change.

11.12.2 Vegetarianism and Substitution

Over 5-8% of Americans declare themselves vegetarians. More than 3.1 million vegetarians in the UK (5.8% of the population. Oddly, some 3 million people in the UK call themselves pescatarians, or 5.7% of the population. Currently, 16% of the UK population follows a meat-free diet (8.6m of the 55m population). They are unlikely to change their diet, but they may if they have a choice with non-animal-derived protein. Many more will if the choice is cheaper food, with no animals killed. Would you prefer to eat a hot dog made with raw meat materials, including muscle trimmings, fatty tissues, head meat, animal feet, animal skin, blood, liver and other slaughter by-products, or just pork? Or chicken. Australians refer to meat pies as “mystery envelopes”, and they may have 20 ingredients. Few components are identifiable, so substitution will be those first. Production of full products may take a decade or two.

11.13 Conclusion Food Disruption

Precision fermentation is here to stay, no matter what political and social pressure is brought to bear. Unlike GMO, where the product is genetically modified, precision fermented protein, fat, or sweetener is indistinguishable from the naturally grown product's protein, fat or sweetener. We often think the change will be whole products - a burger, a steak or something identifiable, but that is unlikely to be the introduction pathway.

Different value chains, regulations, and countries will make timing, order, and species hard to predict. The most likely will be the cheapest. What is clear is that, first up, the substitution of individual ingredients will be the focus. With 44% of all eggs in bakery products, no one will have the information that “fake eggs” are substituting for “real eggs”, or that the “protein” or similar comes from the same supplier. Who will notice if the product is the same, tastes the same, and maybe a little cheaper? The USA, while currently building up anti “fake” messages, will be unable to compete with those 10 companies who provide 75% of their “food” or food products. The list of ingredients on many food products will be reduced as stabilisers, flavours won’t be needed as much in cell-based animal and fish products. The cheaper and more available protein will assist the less affluent population who want more protein, and could relieve the fishing pressure that has devastated marine biodiversity. Food ingredients will be made locally.

The enormous number of animals killed and eaten will be reduced drastically. Almost certainly, chicken will be the first to change, for both eggs and meat. Animals are currently killed just to feed pets, and most owners won't know the difference. Novel products will provide more variety for pets, whether mice or other food sources their ancestors may once have eaten. This change will not affect most of the 550 million family farms, producing 80% of the global food, but the disruption will primarily impact those large super farms. There are not vast numbers but these western farms that make up over 1% of farms over 1,000 ha that take up 40% of the farmland globally.

Vast areas of agricultural land become idle.

- 95% less land for protein production
- 13 times the size of Iowa
- 6 x the size of Germany
- 1/5th of China

What happens when that land is no longer needed for growing crops? By 2035 it could be 50% less than today. We saw that the land used for biofuel production will also become available. The good news is that

- Without crops, that land can become rewilded!
- Rivers and river systems, polluted by chicken faeces and toxic levels of nitrates, can recover from decades of pollution.
- No more plundering of fisheries. Oceans begin to recover.
- With a superabundance of energy and labour costs approaching zero, local production of food within cities makes for a more resilient food supply. Crop loss due to floods, droughts, fire, insects and diseases becomes controllable. Food is local, and optimally nutritious.
- Land, once grazed, can be used for wind turbines and allowed to revegetate.
- Land becomes a carbon sink.
- Within a few decades, CO_2 levels will drop back to pre-industrial levels.

The combination of more

Conclusion of Food:

- The Western diet is dominated by just 12 global food corporations providing food and food products to a billion people.
- The Global south has over 600 million farms that produce food for 5 out of 6 people, over 6 billon, mostly food.
- As countries become more urbanised, with higher GDP and standard of living, Precision Fermentation can solve the problem of unlimited demand for protein.
- Protein production is land-intensive. Beef takes over 120 m^2 to produce 1,000 kilocalories versus chicken at 6.6 m^2 and rice at just 0.7 m^2.
- Protein production is very inefficient. A chicken takes 9 calories to produce 1 calorie of chicken meat. A cow requires 40 calories of food to produce 1 calorie of beef. Precision fermentation is between 10 and 50 times more efficient.
- Precision fermentation will begin the segments where consumers are unaware – in food products.
- Within 2 decades, with precision fermentation the dairy and meat industry will be 10% of the current size and 80% of farmland will be returned to grassland and forest.

---

12 Population

Understanding Population Metrics:

- What is Carrying Capacity? How many people are too many?
- Birth rates and survival rates leading to migration?
- How will increasing longevity due to better health and nutrition determine our future society
- How much land do humans need?
- Inequality. Why is it increasing? What will be the consequences?
- When will the global population peak, and what happens after it does?

Concerns about the number of people the Earth can support have re-emerged recently. Defining an upper limit on the Earth's population is inevitably uncertain. Most published estimates suggest a range of 8 to 16 billion people – a range we are entering. Many think we are overpopulated given the current climate crisis; the destruction of habitats and species indicates to many that the Earth's capacity is less than 8 billion people. [124] We must shift our focus to view the Earth as a system rather than a single number. Instead of one measure, population, we need an ecological approach. Everything is connected. We know:

- Humans are a system. In the health chapter, we discuss why we need to consider our total system, including microbiota.
- As a civilisation, we operate in a complex modern system of communities, ranging from families to local societies to nations.
- Our organisations and businesses are systems.
- Our communities and our nations are systems.

- We live on Earth as a system. Trees, soils, oceans and all biological systems are part of a sustainable planet.

All these systems are interconnected. What is also clear is that we have screwed up our biological systems. Between 1970 and 2016 we have lost over 68% of land vertebrate systems and over half of forests. The cause is agriculture and animal farming. The oceans have fared worse, with a 66% decline in predatory fish biomass over the last century, accelerating significantly over the last 40 years.

> *The world has enough for everyone's need, but not enough for everyone's greed.*
> Mahatma Gandhi

Earth's population is a contentious issue, and we will explore it from different perspectives. The concept of "*carrying capacity*" is incredibly complex, and people hold well-established beliefs that are not based on science but on political views. The world is not equitable. Neither is it sustainable in its current form, as the loss of land biodiversity and destruction of marine biodiversity are a testament to this. We have screwed our home. How do we fix it? Do we reduce the number of people, or do something meaningful about our biological systems? We will look at what we do moving forward and what options we have with regards to the carrying capacity:

1. What is the sustainable human population?
2. What is a sustainable resource demand?
3. How do we restore biological diversity and the environment?
4. Land area needed.
5. Food and energy measures.
6. Financial measures – GDP and Wealth.
7. Inequality.
8. Political and tribal behaviours.

We will explore the population over the last 100,000 years, the anticipated population peak, birth rates, life expectancy, childhood mortality, and other factors that we have some control over.

12.1Carrying Capacity

In 1968, Paul Erlich published *The Population Bomb* [168], predicting worldwide famine and mass death due to overpopulation. It is worth considering how zero population growth works. Erlich's predictions didn't pan out, but his ideas launched a debate still raging today.

In 1972, the Club of Rome published "*Limits to Growth*" and concluded that the Earth, given population growth, food production, industrialisation, and environmental pollution, was bound to collapse unless behaviours changed. The warnings of imminent ecological disaster led environmentalists to believe that an environmentally sound society had to be a "no-growth society," and that achieving it had to become a priority everywhere in the world. The Club predicted the birth rate would continue to climb exponentially, and we would eat ourselves out of house and home. The study was first updated in 1992, then 2012, and most recently, fifty years on from the first version, in 2022. Wrongly, the first study underestimated the reserves of raw materials, but subsequent revisions have shown that demand changes supply. Their revised projections [169] now focus on sustainability. The "Green" movement adopted a no-growth strategy and still exhibits considerable antagonism toward any form of growth, even in the transition to renewable energy or intensive agriculture.

> *The Stone Age did not end due to a shortage of stones. The Bronze Age provided better technology.*

So too with food, energy, and shelter. The Green Revolution alleviated the predicted food shortage. Renewable energy removes the constraint on oil and fossil fuel supplies. Resource supply and demand change unexpectedly.

A key element of the population bomb is carrying capacity. Carrying capacity means all of these need to be in balance.

- *equity* for all – of being fair and impartial.
- *sustainability* for our society – economic, environmental, and social.
- *functions* effectively.
- *Includes* all life – not just humans.

Unless all are met, we place tremendous stress on Earth and each other, and we will eventually bring about self-inflicted demise. It cannot be short-term but needs to be long-term. Reviewing history and the rise and fall of civilisations, we see what happens when those civilisations fail to function effectively. The total weight of all 8 billion human beings on Earth is about 350 billion kg. That's more than 100 times the biomass of any large animal that's ever walked the Earth, says Wilson in his book *Social Conquest of Earth*. [170]

Are we past carrying capacity? Should we go back to 0.5 billion, the number we had before the 20th Century? Would you be willing to volunteer for the reduction? Or could we be as many as 16 or 40 billion? If it's 2 billion, how do we adjust to get back to 2 billion from 8? Or to 40 billion? Let's examine some of these factors. Keep in mind, Ehrlich emphasised equity, sustainability and that the world is functioning properly. As we have seen with the loss of biodiversity and biomass on land and in the ocean, Earth is not currently sustainable. Either we reduce people, reduce resources used or increase the biomass. How?

12.2 Earth Population by Year

Earth's population is currently about 8.2 billion. Actual numbers are no more than guesses as many countries have poor governance. Humans have had a couple of close calls. A genetic study called Fitcoal suggests [171] just 1,280 survived at 813,000 to 930,000 years ago in the mid-Pleistocene, with 99% wiped out, and a more recent event called the Toba catastrophe at 70,000 years ago dropped the population to as low as a thousand to ten thousand people. That is a close call!

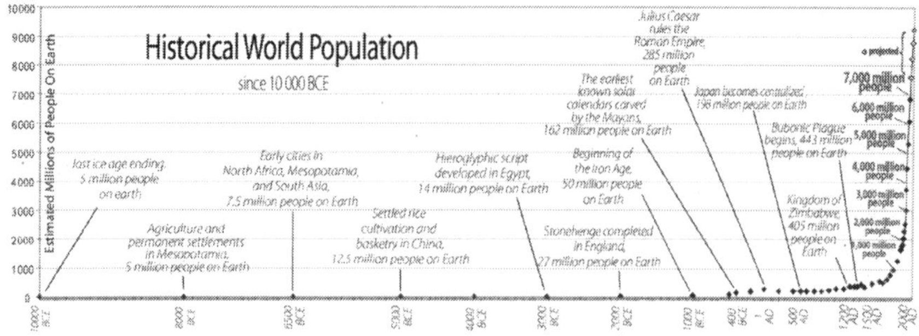

Figure 1: World population began accelerating about 4 000 years ago and has "exploded" in the past 1 000 years. *Source: UNEP-GRID Sioux Falls; population data – US Census Bureau 2011.*

Figure 42 Historical world population.

Figure 42 shows how the historical world population since 10,000 BCE remained about 4 million at the end of the last ice age. By 5,000 BCE during the rise of China, the population was only about 12 million. The Iron Age saw about 50 million, and Caesar's world was 285 million. Since then, the population has ballooned. About 900 million in 1800, 1.65 billion by 1900, and we rushed past 2 billion by 1928 [124]. China and Astrophysicists Adam Frank and Gavin Schmidt of the NASA Goddard Center proposed the "Silurian Hypothesis" in 2018, suggesting an advanced civilisation existed before the Carboniferous geological record, 350 million years ago, or even 12 million years ago. The name Silurian comes from the Dr Who TV series. [172] This intriguing theory, along with other pre-*Homo sapiens* theories, is met with scepticism and debate within the scientific community.

We will tease out 5 factors contributing to that population increase from 4 million to 8 billion. Birth rates, infant mortality, and living longer all contribute to population, and multiple ways these are influenced. Now, most predictions show we are close to peak population and won't reach 10 billion. [124]

12.2.1 Population Peaks in 2040

Depending on the model, peak population will be somewhere between 2040 to 2060.

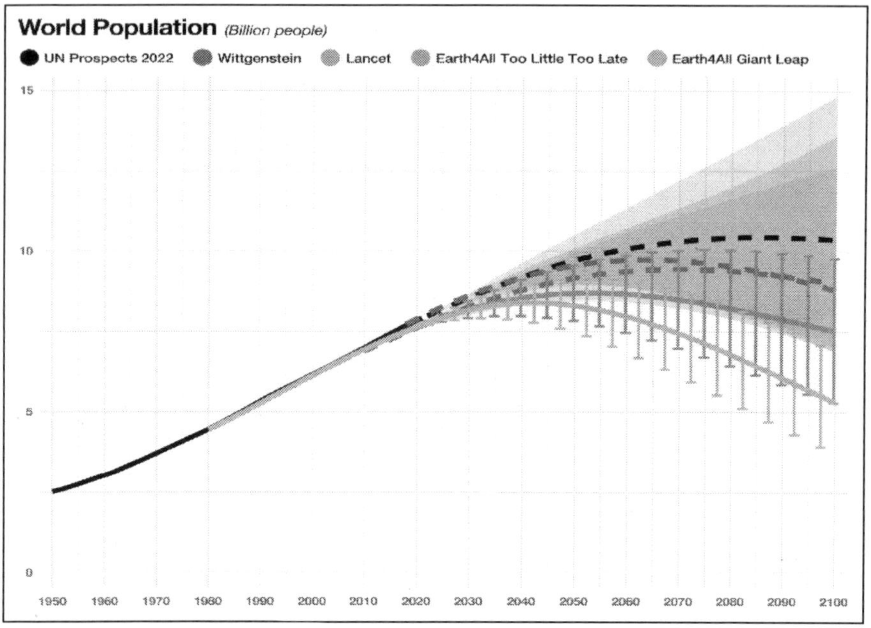

Figure 43 World population in 2100.

Figure 43 shows that the peak population will be earlier than 2100. Most models predict the population will be lower in 2100 than it is today; some even predict 35% lower, at 5.5 billion. The UN median fertility model, the IISA model, the IHME model and the Lancet model vary substantially about when the peak occurs and the numbers in 2100.[173] The Club of Rome equality model says that if the world invested more in education and we had less inequality, numbers would be even lower.

Short of a drastic decline in fertility, there will be more people on Earth at the end of this century than there were at the beginning. About the same as today.

12.2.2 Birth Rates

For society to be in a steady-state stage, the average number of births per woman needs to be about 2.1, and we now see that most countries are below that. The global total fertility rate (TFR) has halved in the past 70 years, and rates have continued to fall over the past 5 years. Korea is now down at 0.8, China 1.3 and the US 1.8. Education, especially of women, is a key driver for lower child numbers.

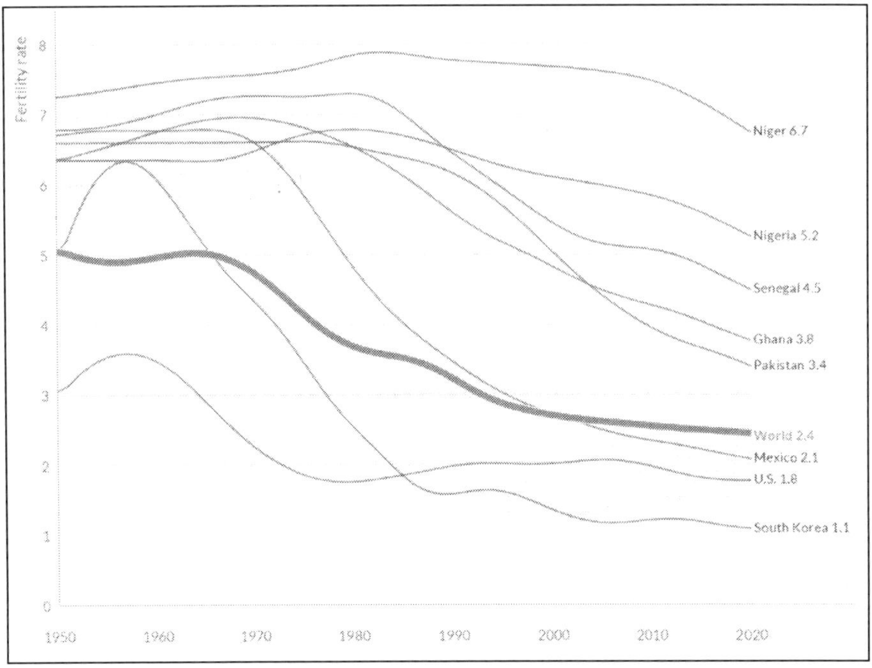

Figure 44 Plummeting world fertility rate across the world.

By 2050, over three-quarters (155 of 204) of countries will not have high enough fertility rates to sustain population size over time; this will increase to 97% of countries (198 of 204) by 2100. Government intervention makes little dent in the falling birth rates. Australia had a treasurer who said have one child for mum, one for dad and one for Australia. South Korea spent over $3 billion to increase birth rate with zero impact. The USA recently promised the pro-natalist movement $5,000 per child although that is chump change when it takes over $237,000 to raise a child. Comprehensive social programs and immigration reform could help, but those are almost as politically toxic. Russia has a population of 143 million with a birth rate of 1.4, the lowest since the 1700s. It could fall to 80 million by 2100 and the regime is restricting access to abortion and contraception and banning "childfree ideology". It calls the situation catastrophic, not least due to the estimated 200,000 to 500,000 dead due to the invasion of Ukraine.

Globally, we had high birth rates in the 1960s and 1970s, along with a rapid increase in life expectancy from 30 to 60 years, which caused a significant population growth. During that time, birth rates per household remained high.

By the 1990's education, contraception and social changes brought birth rates below replacement levels. In many ways, declining fertility rates are a success story, reflecting not only improved access to contraception but also women's freedom of choice to delay or have fewer children, along with increased opportunities for education and employment. [128]

According to recent projections, only 8 per cent of the world's births in 2026 will occur in Europe, North America and Oceania combined. In absolute numbers, that translates into roughly 4 million births in North America, 5 million in Europe and fewer than 1 per cent of global births in Oceania. The overwhelming majority of humanity's newest members, 85%, will instead be born in Africa and Asia. Asia alone is expected to account for approximately 65 million births in 2026, representing about 49% of all global births. Africa will follow, with more than 47 million births, accounting for 36% of the total. Latin America and the Caribbean will account for 7%. [174]

12.2.3 Increase in Life Expectancy

While fertility rates are falling, increases in life expectancy [175] contribute to uncertainty about the population peak. Most countries are heading towards life expectancy over 80 years (currently 73), and it is this rise, not fertility rates, that is driving global population increase. Conspiracy theorists and racists with their replacement theory misunderstand population science. Rising life expectancy has major economic and health implications. Increased illness and a rise in dementia in many countries challenge health budgets. Old people do not buy most goods and services or participate in the workforce. They are a significant drain on health services.

Region	1950	2000	2050F
North America	68	77	84
Latin America & Caribbean	49	71	81
Europe	63	74	84
Asia	42	68	80
Africa	38	53	68
Australia & Oceania	61	75	82

Figure 45 Life expectancy 1950, 2002 and forecast 2050 by continent.

Figure 45 shows most countries are heading towards an 80-plus life expectancy by 2050. Notable increases are Latin America (50 to 81), Europe (63 to 84), and Asia (42 to 80). Even Africa is rising from 38 to 68.

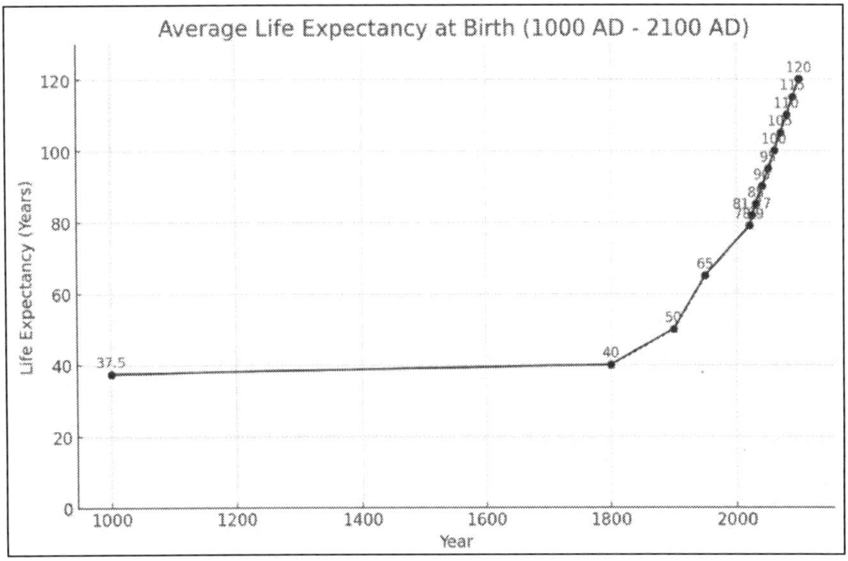

Figure 46 Increasing longevity over past 1000 years

Figure 46 shows the dramatic change in life expectancy over the past 1,000 years. The dramatic rise since 1800 and the continued rise now have been driven by 2 factors: reduced child mortality and better health care, as we will explore more. Nowhere is it more stark than in Australia. Over the 20 years to 2018, the number of nonagenarians (those aged 90-99) grew from 63,000 to 181,000. Life expectancy is projected to rise by a further 5 years by 2050, to 82 for males and 87 for females. The number of Australians aged 85 years and over is projected to double by 2042. In the 100+ group or centenarians, numbers have gone from 1,300 in 2000, 3,700 in 2010, 6,000 in 2018, and are projected to be over 50,000 by 2050. External risks (meteors) or internal risks (nuclear war, earth catastrophes, disease, climate change) could change the population overnight. The SARS-CoV-2 pandemic dropped life expectancy in the USA and other countries by a couple of years as millions of older people died, and a repeat pandemic is possible. Advances in health could increase life expectancy by decades, for those who can afford them—more of that in the health chapter.

12.2.4 Childhood Mortality

Childhood mortality was the major driver of the increase in life expectancy.

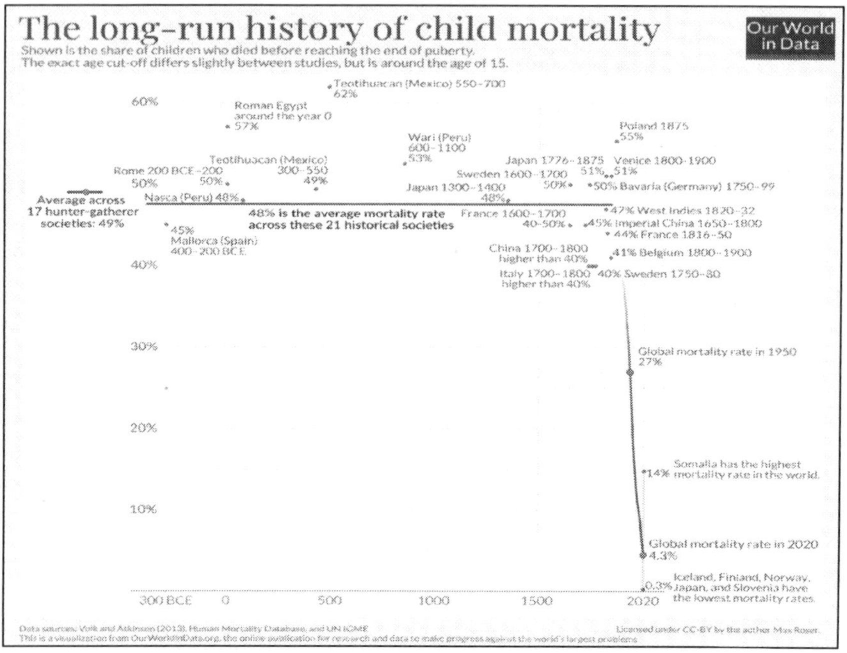

Figure 47 Child Mortality over past 2 thousand years

Figure 47 from Our World in Data [176] points to the rapid rise in population in the 1950s, and children went from assets (working on the farm or caring for parents in old age) to liabilities. The role of children looking after their parents in old age changed when social welfare programs took over that role in most Western societies. No longer were children required to support their parents. The left side, with ancient and early historical societies (300 BCE – 700 CE), shows the highest recorded child mortality rates, including those from Teotihuacan and Roman Egypt, alongside the hunter-gatherer baseline.

- Hunter-Gatherer Societies: 49%
- Roman Egypt (Around Year 0): 57%
- Teotihuacán, México (550–700 CE) 62%

In the middle section, in the medieval and Pre-Industrial Era (600 CE – 1800 CE), which hovered around a 48% historical baseline.

- Wari, Peru (600–1100 CE): 53%
- Sweden (1600–1700): 50%
- Bavaria, Germany (1750–1799): 50%
- France (1600–1700): 40%–50%

Lastly, on the right we see the Industrial Era and the Modern Decline (1800 CE – 2020). We had the tail end of historically high rates in the 19th century, just before public health advancements triggered the massive, steep plunge toward modern global lows.

- Poland (1875): 55%
- Global Rate (1950): 27%
- Global Rate (2020): 4.3%
- Somalia 14%.
- Iceland / Finland / Norway / Japan: 0.3%

Child mortality plummeted, even in Somalia, which has the highest global rate. That's one-third of the 50% for the past 2 millennia. The decrease in infant mortality is largely due to antibiotics, vaccinations, sanitation and urbanisation. Nowhere is the relationship stark when we look at child mortality versus a country's income or GDP.

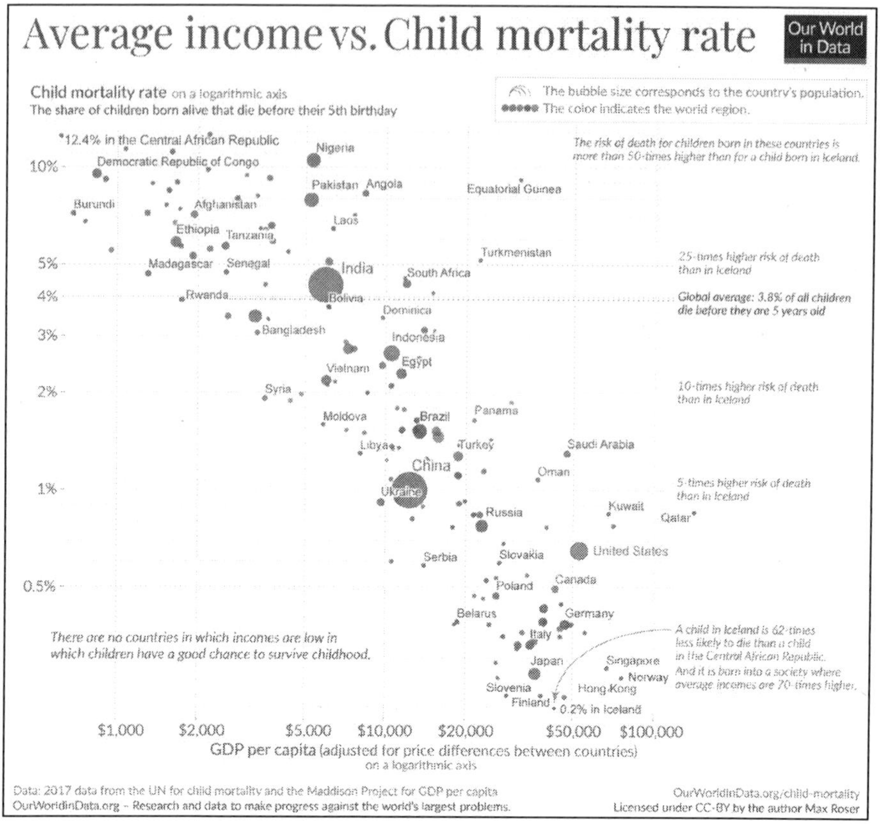

Figure 48 Average income versus Child mortality

Figure 48 [177] shows the log of the inverse relationship between average income and child mortality rate. With a GDP of $1,000, mortality is 12.4% in the Central African Republic, compared with a global average of 3.8% in countries such as Rwanda and South Africa, and as low as 0.2% in Iceland. The child in Iceland is 62 times less likely to die than in the Central African Republic. There is no country in which incomes are low, in which children have a good chance of surviving childhood. It is clear that the driving influence is not biology but wealth inequity, and that is driven by power and greed. We will deeper into inequality later in this chapter.

12.2.5 Interaction Between Factors

Interactions among birth rates, childhood mortality, economics, social cohesion, longer lifespans, health programs, poverty, war, famine, climate change, and family size are complex. We are currently at a place where there are no effective levers to pull. Government programs to increase birth rates are ineffective. The African continent is struggling with health programs due to ongoing armed conflict. Given the declining birth rate, improved health programs, and rising life expectancy, the model of peaking in mid-2050 and then falling to under 8 billion is the most likely. Population is mostly baked in. We don't have any levers to pull for the human population, so, as we will see, we have to rely on changing our resource use and agriculture to bring the Earth back onto a sustainable trajectory.

Pro-natalism (the idea that people should have more babies than they choose to have at present) has become an established orthodoxy. [178] The central claim is that unless something changes soon, human populations, both global and national, will decline rapidly, with many negative consequences. This is simply not true, based on our understanding of fertility and life expectancy. In fact, if the global birth rate were down to 1.5, we would still have a billion people on the planet by 2100, over 300 years on from Thomas Malthus's *An Essay on the Principle of Population*. [179] That book, first published in 1798, argued that human populations grow exponentially (geometric progression), whereas agricultural resources grow only linearly (arithmetic progression).

12.3 Longevity Escape Velocity

Futurists point to a concept called "longevity escape velocity". The idea is that as our life extension technology improves, our life expectancy could increase by more than we age over a set period. For example, as medical innovations continue to advance, we would still age a year at a time. But if our life expectancy were to increase by a year and two months, we would effectively gain two months of life. Ray Kurzweil—former Google engineer and prominent AI-centric futurist—told multiple outlets he believed humanity would achieve longevity escape velocity by 2029.[180] He pointed to the development of the COVID-19 vaccine that took just 2 days. Not years. Aging isn't just about growing older as time passes; it's about how fast you're aging. The speed of aging refers to how quickly your biological systems accumulate damage and decline in function. Two people born on the same day can have drastically different biological ages depending on their lifestyles, genetics, and environments. According to statistical projections, even a 30% increase in remaining life expectancy would mean 20 extra years. This additional time would likely yield further advancements in rejuvenation technologies, continuing the cycle ad infinitum.

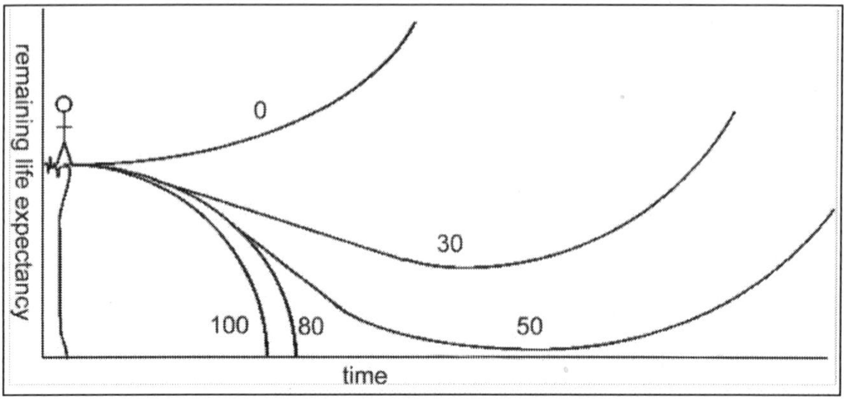

Figure 49 Longevity Escape Velocity based on your current age.

In Figure 49, if you are 100 today, you would have no LEV, as you have run out of time. If you were born this year, you could not predict your life span, and in fact with every passing year, your life expectancy would increase. With this longevity escape velocity in play, how long could humans live? 120, 150 years? The bowhead whale lives 211 years and the rough-eyed rockfish over 200.

Achieving longevity escape velocity doesn't guarantee you will live forever, as you must live well and avoid accidents. We have to live with the unpredictability of life. We will not cure all cancers in the next five years, and cancer is a disease defined by the randomness of the mutations that cause it. Technological advances like self-driving cars might reduce the odds of a deadly accident. Eliminating fossil fuels and pollution from vehicles would stop 10,000 Australians dying each year of asthma. Note that the existence of medical treatments and advances is not synonymous with their widespread implementation. Tracey will be a beneficiary; Ada won't.

12.4 Land Needed for People

If all 8.2 billion people lived in cities, how much land would be needed? Reuters fact-check [183] debunks posts claiming the world's population would fit within a 50.8-mile-by-50.8-mile square. This density lacks water, sewage, energy, housing, recreation and transportation. A more realistic view of the space needed would be to use a large city such as Paris. Greater Paris, Métropole du

Grand Paris (MGP), is densely populated and almost completely urbanised. Some areas are more densely populated than others. Data [181] show the central area is the most densely populated, with over 500 inhabitants per hectare (50,000 per km^2), but with lower density, down to 50 people per hectare (5,000 per km^2), in the south-east areas of Territoires Grand Paris Sud Est Avenir and Grand-Orly Seine Bièvre.

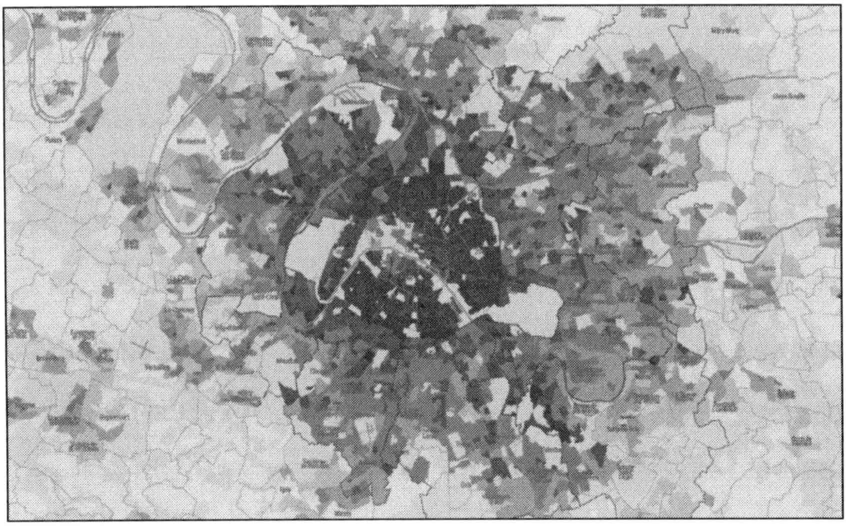

Figure 50 Population density of Paris - a city of 7m people

Some cities are higher density than Paris with [182] Beijing 14,000 people per km^2, Shanghai 16,982, Hong Kong 6,700, Chicago 4,600, New York 11,300 and San Francisco 7,200. At Paris's average population density of 20,641 inhabitants per km2, the whole of the world's population would need 400,000 km^2 which is a square measuring 630 x 630 km. An area of 400,000 km^2 is smaller than Texas, which is 670,999 km^2. Countries of about 400,000 km^2 include Germany (population 83 million), Sweden, Italy, Spain and Morocco. The mega-cities of Guangzhou, China: 73.6 million; Shanghai, China: 42.3 million; Tokyo, Japan: 41.3 million; Delhi, India: 36.9 million; and Jakarta, Indonesia: 30 million, all function effectively. There is nothing to suggest that other cities would not function effectively. They have transport, food, and waste all sorted. We could have more people if we were to accept living in a city like Paris. It is clear we could be much more than 8 billion. Would we want to or need to?

12.5 Food Supply and Inequality

Walter Malthus (1800s), with his Malthusianism theory, is that population growth is potentially exponential, while the growth of the food supply or other resources is linear. This eventually reduces living standards to the point of triggering a population decline, as popularised by the Club of Rome. As with energy and shelter, technological improvements can overcome perceived constraints.

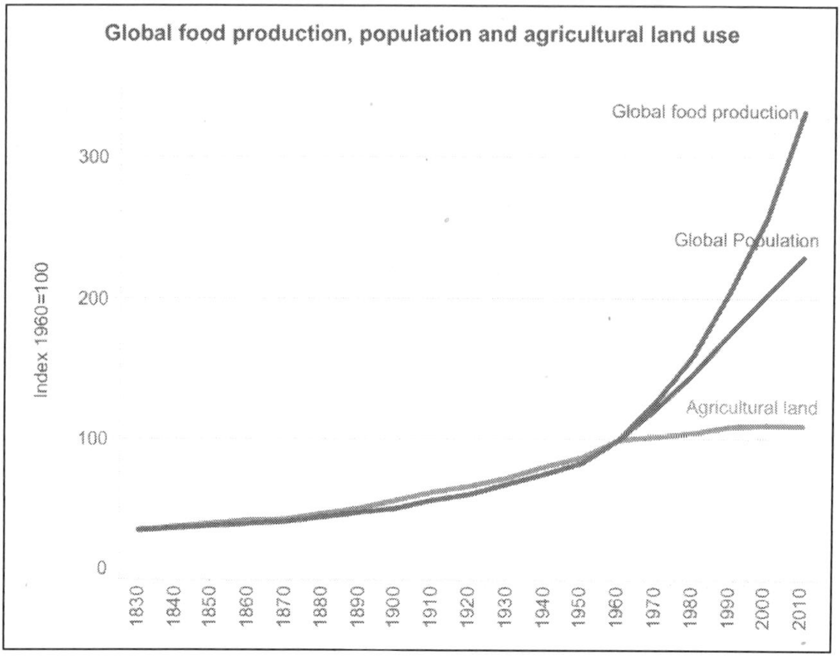

Figure 51 Global food production trebled from 1960 to 2000

Figure 51 shows divergence beginning about 1960. Subsequently, agricultural land has increased by less than 10 per cent, all the while the population has increased 2.3 times. Food production has more than tripled. Farmers and fishers have consistently increased production, outpacing population and income growth. Despite this, seldom have either the resources or the education to access the latest technologies, and often hesitate to try new approaches they perceive as too risky to their own food security and livelihoods. As we have seen, and the OECD states [183], we have this marked difference between the Western countries and the global South.

- In advanced countries, technological innovations bring high-tech solutions.
- In the South, those 550 million family farms with less than two hectares represent 84% of farms globally, produce 80% of food but only make up ⅓ of global crop production.

The difference is that in the small farms, they focus on plants. On corporate farms, they produce crops for feeding to animals or for burning in cars.

12.6 Green Revolution

The concept of food insecurity was dispelled with the Green Revolution. Norman Borlaug [184] is credited with being the father of the Green Revolution in the sixties and seventies, with doubling wheat yields in India, Pakistan, and other countries. New types of wheat in Mexico and new types of rice in India yielded more than we ever had. Along with drought-resistant, flood-resistant, pest and disease-resistant plants, the "carrying capacity" soared.

Since the 1950s, we have had a dramatic fall in the number of malnourished people. Numbers have dropped from 65% in 1950, 25% in 1970, 15% in 2000 and 9% in 2020. Remarkably, over these 70 years, the global population rose from 2.5 billion to 8 billion, so why is it that 0.8 billion people go to bed hungry at night? More than 30% of food grown feeds animals. We have enough food to feed those 0.8 billion. It is inequality and politics that prevent it. Perversely, more die from obesity and lifestyle diseases than from hunger today.

12.7 Energy Supply and Inequality

We looked at those 0.8 billion, like Ada, in Optimistic Storm: Part 1 in the chapter on Energy Poverty. We could change the lives of those who live in energy poverty, deprived of heating, cooking, lighting and water, and we pointed out that for less than $1,000 each or $300 billion over three years, it can be solved. $300 billion is the expected cost of 6 nuclear-powered submarines for Australia to defend trade routes. As we saw in energy, bringing energy to Australia will be replaced with renewable energy. Overseas manufacturing will be disrupted by humanoid robots. Which trade routes? As Forrest Gump said, "Stupid is as Stupid Does". Can't help stupid.

The world has inequality. Food, energy, health and wealth. The inequality in energy includes:

- 0.8 billion people in energy poverty.
- One billion people use over 70% of global energy.
- One billion people use just 10% of global energy.
- Burning too much fossil fuel has driven human-caused or Anthropogenic climate change to long-term changes in Earth's weather patterns and temperatures.
- Exceeding 1.5°C will result in a billion climate refugees.
- We face a 30% reduction in GDP due to climate change.

Global primary energy use is about 166,000 TWh annually. Fossil fuel is very inefficient and wasteful with over 60% of primary energy wasted for extraction, transport and use. Technologies exist today that are economically viable for increasing electricity generation from 30,000 TWh to over 66,000 TWh. That would meet the forecast global energy demands. Transport energy efficiency rises from under 20% to over 85% as we transition from internal combustion engines to battery-electric vehicles. Gas heating is about 85% efficient, but replacing that with heat pumps is over 300% efficient. Canadians each use over 110 MWh each per year, whereas sub-Saharan Africans use less than 0.1 MWh each. That is about the amount a Canadian uses to power a refrigerator for a month.

The Jevons Paradox, proposed by economist William Stanley Jevons in 1865, suggests that technological advancements that increase resource use efficiency can paradoxically lead to an increase in overall resource consumption. This occurs because lower resource costs due to efficiency lead to increased usage, potentially outweighing the efficiency gains. Given that the amount of mining required under a renewable energy economy is at least 580 times less than current mining, there is some flexibility. Furthermore, as recyclable technologies develop, there will be a growing demand for additional resources over the next 20 years. There are over 1.4 billion vehicles with utilisation of less than 4% and most are internal combustion engines. Personal public transport (robotaxi technology) will operate at over 40% utilisation and be 5 times more energy efficient per km. That means we would only need 150 million cars, less steel, and fewer resources for equivalent transport.

The renewable energy regime is neither polluting nor as energy intensive. The total energy needed globally is equivalent to about 1.5 hours of sunlight capture. We might need 2.5 hours of sunlight if we increase energy use. The extra energy demand with AI, and from those 800 million energy-poor people, would not exceed what Canadians use today.

12.8 GDP and Wealth

While food, energy and land are physical, economic inequality is the real problem that we and the political system appear incapable of solving. Wealth inequality is the real challenge. That is the most likely cause of civil unrest over the next few decades.

- One billion people consume 71% of all resources, with an income greater than $100 per day.
- A billion people consume less than 1% of resources, living on less than $1 daily.
- Six billion consume 28%, living on $2 to $100 per day.
- Less than 1% (10 million people) have 43% of global financial assets.

12.8.1 Billionaires

Billionaires. This select group of 3,500 billionaires own more than $15 trillion, [185] which is over 5.4% of global GDP. The number of billionaires is rising by 10% per year, while their wealth has increased by 13.4% in the past year. There are billionaires, and they become even wealthier. Were you one of them? Some analysts suggest the number may be under-reported, as many keep their wealth private. These billionaires possess wealth that is almost beyond comprehension. Just 400 billionaires hold 50% of America's wealth. Or 1% of the wealth of the world's billionaires would be sufficient investment to eliminate energy poverty. In the USA, 3 million Americans make up the wealthiest 1%, and they collectively are worth more than the 290 million who make up the bottom 90%.

Australia, with its 178 billionaires, has the most billionaires per capita. An analysis shows there is a broad range of self-made versus inherited billionaires. On average, 67% of billionaires are self-made, and 33% are inherited.

12.8.2 What Causes Inequality?

This book won't argue the causes and doctrines of inequality. In some countries, politics and war have kept this gulf between the rich and the poor. For many countries, a small number of people or families control their country. In Saudi Arabia, just 2,000 people control most of the country. The Philippines has about 319 family dynasties with fewer than 200 who control politics and wealth. Globally, over 10% of all leaders come from family dynasties.

In the West, absurd doctrines, such as "trickle-down economics" and Keynesian theories, have perpetuated inequality. Free market Reaganomics, Thatcherism, and Rogernomics privatised government institutions. Globalisation in the UK (Tony Blair) and the USA (Bill Clinton) embraced a neoliberal approach to trade. These policies assumed that free markets boost productivity and competition, lower prices, and self-regulate more efficiently than administrators. This mindset plays out with corporations. CEOs were paid about forty-two times as much as the average employee in 1980. Forty years on, in 2016, they were paid 347 times as much.

It is ironic that in the 18th century, serfs lived in servitude to a few landholders. Today, serfs (us) are employed by a few techno or mining billionaires. Large estate houses have been replaced by superyachts and mansions. It is not entirely negative. In 1980, 43% of the world lived in extreme poverty (as defined by the World Bank), whereas today that number is approximately 8%. Globalisation has lifted a billion people out of poverty in just forty years, yet a billion remain impoverished.

12.8.3 Rent Vs Tax

One view of this inequality is that it involves paying rent, not taxes. The big difference is we now pay "rent" to corporations and previously we paid taxes to the government for provision of services. In the 18th century, during the age of serfdom, we paid the Lord of the Manor. Nowadays, the entry fee exceeds the cost of the services received. For example, the water bill that was once public is now private and has doubled or tripled in price since the government sold the services. Rent is a kind of private tax that the poor and middle class pay to the rich. Many times, it is extortion, and the payments we must make when the rich have us over a barrel. Food. Interest, electricity, water, health, railways, education and in many countries, high housing rentals. No matter which country, the rich have cornered services and facilities, and people pay that tax to individuals or corporations, not to the government. Many who dislike paying taxes to the government fail to see that the alternative is taxes to the modern capitalists. The widely held view of "Free Markets" was government-owned utilities were inefficient, and that corporations were efficient. That myth has been well and truly debunked and few governments have the political stomach to nationalise previously floated service providers. Health for example, costs 3 times more in the USA than in countries with government health care for equivalent outcomes.

We cannot change this political system easily. Politicians and business have a cosy relationship, often with a revolving door between government and business. The Game of Mates book has a good analysis from an Australian perspective [186] and it details how hundreds of billions of dollars a year move from the public service sector to the business sector. We should not be optimistic that the drivers for inequality in these systems will change anytime soon. While people are getting wealthier as measured by GDP, health, leisure and education, many feel they are worse off and blame governments of all persuasions.

12.8.4 Wealth Disparity

The global wealth pyramid [187] shows how skewed wealth distribution is.

- People with over $1 million hold more than 45% of the $454 trillion global wealth.
- Most homeowners in the Western world have million-dollar houses. Where do you fit?
- The top 1% of households hold 32.3% of the country's wealth, while the bottom 50% just 2.6%.
- The US dollar has lost over 96% of its purchasing power since the creation of the Federal Reserve in 1913. A company with 31 million in 1913 would have the spending power of a billionaire today.
- People are getting wealthier, even if they don't feel it.
- Millionaires have increased from 0.4% to 1.1% in the past 20 years, and over 1.5% by 2027 with over 86 million millionaires.

The trend is clear. The rich are getting richer and the poor are getting poorer. America's super-rich have grown about $21 trillion richer since 1989. The poorest 50% by contrast, have become $0.9 trillion poorer.

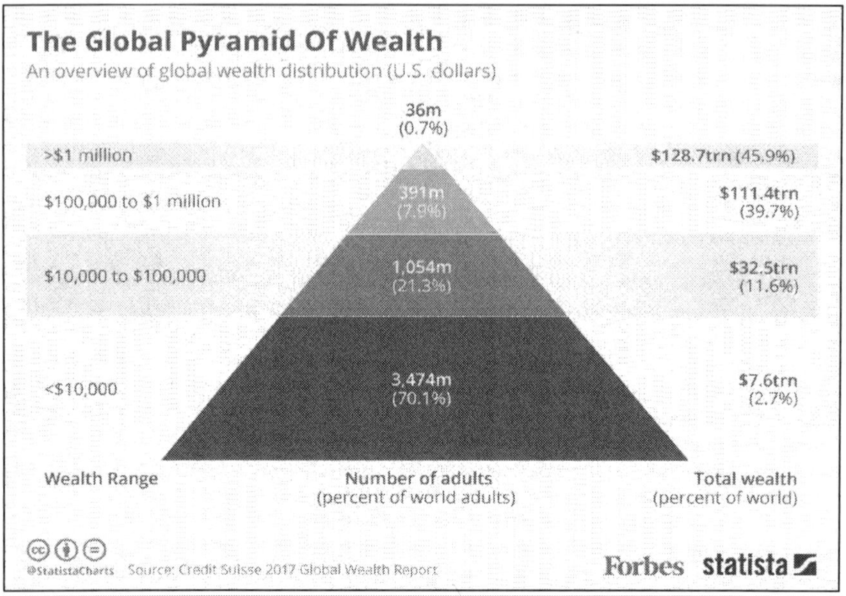

Figure 52 Share of wealth globally shows the wealth is concentrated.

Figure 52 shows this global wealth pyramid, but it does change over time. Many Western countries have followed that trend of the rich getting richer and the poor getting poorer. In contrast, some countries, including China, have brought 800 million out of poverty. In the table below, we see that the share with more than $1 million has increased from 0.4% twenty years ago to over 1.5%. Those with less than $10,000 has halved from 80% down to 46%.

Net Worth Global Adults	2000	2022	2027P
More than $1 million	0.4%	1.1%	1.5%
$100,000 to $1 million	5.5%	12.0%	14.8%
$10,000 to $100,000	13.4%	34.4%	37.0%
Less than $10,000	80.7%	52.5%	46.6%

Figure 53 Share of global wealth increasing to 2027

A measure of wealth may be the amount of steel per capita. The USA has 10 tonnes of steel per person. In Nigeria, it is 1. Steel is used in cars, public transport, hospitals, schools, and infrastructure, often in combination with cement. The carbon emissions from steel (7%) and cement (8%) account for a sizable share of emissions in countries with the most resources.

12.9 Conclusion of Population

We have introduced the concept of carrying capacity, defined as a sustainable and equitable functioning society. Food, energy, materials, and environment are all constraints. Most everything we use is harvested or mined. Billions of tonnes per year. From clothes, wooden chopsticks, pencils, electronic equipment and all the roads, buildings and infrastructure - it is either harvested or mined. Since the dawn of humankind, we have utilised our environment to create tools that enable a better lifestyle. Our tools may have been fashioned from the animals we hunted - skins for clothing to keep us warm, bone tools to make hunting or gathering food more effective, and even mining for gold and silver. There is no doubt that we are exceeding our carrying capacity today, given what we have done to other living creatures and our environment. We have screwed up our planet and are living beyond the Earth's carrying capacity. What do we do now?

"Business as Usual" is Unsustainable for the following reasons.

- Climate change is worsening. [188] Every measure and prediction of human-induced warming is now over 1.36°C, and the rate of warming is now over 0.3°C per decade. Huge consequences include a 3°C change that will cause a recession of at least 30% - like the Great Depression, but forever. A billion climate refugees. Rising sea levels. Worse storms and fires.
- Population. We have few socially acceptable levers to control the population. We don't even know exactly how many people there are on the planet.
- Birth rates have fallen below replacement levels in most developed countries.
- Education and improved health for those who still have high birth rates will help, but we must avoid usurping their sovereignty.
- Life expectancy is increasing. Who would not want better health?

- The population may increase from our current 8.1 billion up closer to 9 billion, but it will almost certainly fall to under 7 billion by 2100.
- Inequality of resources. One billion people consume 70 times as much as the poorest 1%, and there is little appetite for change.
- Energy shortage. We face energy shortages and disparities in energy access, but the transition to the 4th era of electricity will create a superabundance of electricity. By 2040, we will have entered that era, and energy poverty will be a thing of the past.
- Emissions. Burning too much fossil fuel has driven human-caused, anthropogenic climate change but we know that we can stop emissions with existing technology. But do we have political will?
- Food disparity. While the number of obese people is now greater than the number of starving people, the question is why. Food production and agricultural land have increased at a faster rate than the global population. But we feed animals, not humans. We burn food. We have enough food.
- Species loss. Human activity has led to a 68% decline in vertebrate species between 1970 and 2016. Something needs to change, yet politics and human pressure for food and materials highlight our unsustainable practices. Countries continue to send out ships to further deplete ocean biomass.
- Loss of habitat. Since 1970, half of the world's forests have been lost to agriculture and livestock farming.
- Freshwater is running out and will be a crisis by 2030.

We have limited options for humans and our environment.

- Reduce the number of humans. Who will volunteer?
- Consume less. Good luck in convincing the wealthy to share more equitably!
- Change our energy supply. Underway but could be faster.
- Change our agricultural systems from meat to plant or cell-based protein.
- Use "free labour" that humanoid robots will enable to replant and repair the environment.
- "Undo" some of the environmental damage.

We do understand the challenges but appear powerless to take action. We have fantastic developments for sustainability and the Earth's population increase is slowing down. The knowledge of the wealthy is now available to the poor (thanks to internet disruption). Over the next 2 decades, technology will provide better health, low-cost energy, clean water, sanitation and education. Is it too slow?

Remaining barriers are politics and religion. We continue to see repressive societies with no simple mechanism to switch them off. In many respects, the political system needs to be sustainable. Countries with political upheavals fail to advance, and we can highlight any number of countries with multiple failed political systems. Citizens pay the price of hunger, environmental degradation, disease, and poverty. Knowing that civilisations throughout history have had equally poor politics is scant comfort for those in those repressive regimes, even though we know there is a better future. Climate change is beginning to add to their misery, and they face the real prospect of worsening climate impacts. And lastly, wealth inequality is the major barrier to fixing these issues. Wealth is power, and the wealthy, as a cohort, have little or no desire to relinquish that power and effect change.

Conclusion of Population:

- Carrying capacity is about equity and sustainability, not just a headcount – and on current evidence we've already overshot it.
- Global population is set to peak around mid-century and then decline – falling birth rates, not famine, are the story.
- Wealth inequality, not resource scarcity, is the real barrier – a billion people consume 70 times what the poorest billion do.
- We have few levers over population itself – the real levers are energy, food systems, and how resources are shared.
- Longer lifespans and lower birth rates will reshape society – ageing populations need different infrastructure, not just more of it.
- How many people the Earth can support is the wrong question – the right one is how we choose to live together

---

13 Health

Introduction to health:

- Genomics and multi-omics: how understanding our biology at the molecular level is transforming medicine.
- Lifestyle diseases – cancer, heart disease, dementia, and Parkinson's – and what drives them.
- The hidden toll of chemicals and plastics on human health.
- Why measuring more – wearables, dashboards, and continuous monitoring – could catch disease before symptoms appear.
- Why so many preventable deaths still happen, and how earlier detection changes the odds.
- Antimicrobial resistance: a quiet health crisis that disproportionately hits the poorest communities.

The UN Universal Declaration of Human Rights, established in 1948, declared that everyone has the right to a standard of living adequate for health and well-being, including medical care. Improvements in health in the 20th century have led to significant increases in life expectancy. Vaccines account for 40% of the decline in infant mortality over the last 50 years, with plummeting rates globally; they've fallen by over two-thirds, from around 10% in 1974 to less than 3% today. That's 154 million children alive today who otherwise would be a statistic. [189]. Over 60% of those millions were due to measles vaccination, and 95% were among children vaccinated before age 5. Which begs the question why in this era, some are challenging the very notion of vaccines?

Much of the transformation of health in the future will depend on AI and the successors to genetics, multi-omics. None of these will help those who don't help themselves. A headline recently said "'I'm so sorry, I've got cancer: why do two of every five Australian men die early from preventable conditions?" A recent study shows men are less likely than women to have a trusting relationship with a doctor and so miss out on health literacy and advice. The consequence is that cancers and heart diseases are often not detected early, and as we will see, in almost all cases, early detection simplifies treatment and reduces mortality. [190] The top five leading causes of years of life lost in men are coronary heart disease, lung cancer, self-harm/suicide, chronic obstructive pulmonary disease (COPD), and stroke. These are largely avoidable through lifestyle and behaviour change, screening and earlier diagnosis. The cost of COPD is an underdiagnosed, undertreated, and yet largely preventable disease and the treatment cost beggars belief. Australia spends $11 billion annually treating the preventable cases of those five conditions. In the USA, COPD costs over $50 billion alone. [191] Suicide and non-fatal harm costs over $500 billion. In the UK, all cancers cost over $20 billion. The cost of treatment in the Western world is staggering, whereas in the Global South, only a fraction of that is spent. The consequences are early death, loss of human life, and lost productivity and societal wealth.

13.1 Genetics

We probably have heard of genomics where the base pairs that make up our RNA, DNA and protein are much like language. The Human Genome Project (HGP) was finally completed in 2003, after 9 months and $100 million. Prior research, spanning 13 years from 1990, was estimated to have cost over $2.7 billion. Advances in sequencing, computing and AI today mean the cost is now down to $200 per test and done in less than 8 hours. The findings of the HGP suggested that all humans are 99.9% genetically identical, and that only 0.01% variation in the human genome can account for phenotypic differences among humans, such as disease susceptibility, drug responses, and physical traits (hair colour, eye colour, height, intelligence, etc.).

The DNA sequence is much simpler than language, with the four bases in DNA adenine (A), cytosine (C), guanine (G), and thymine (T). These bases form specific pairs (A with T, and G with C). RNA is similar except adenine always pairs with uracil (A-U). When it comes to proteins, throughout known life, there are 22 genetically encoded (proteinogenic) amino acids, 20 in the standard genetic code and an additional 2 (selenocysteine and pyrrolysine) that special translation mechanisms can incorporate. Many of us thought DNA, RNA, and proteins were the main focus of study, but how wrong we were as we came to understand biology better. We now have better tools, better measuring machines and better computers and our understanding of health is improving week by week.

13.2 Multi-Omics

Multi-omics may be a new term for many [192] and comprises those words that end in "omics". [193] It is a new approach in which datasets from different omics groups are combined during analysis. The different omic strategies employed during multi-omics are genome, proteome, transcriptome, epigenome, metabolism with lipids and microbiome being added to the term over the last decade. Advances in big data and AI are enabling the understanding of relationships between these different multi-omics rapidly, which was unthinkable even 5 years ago. The human body has about 30 trillion (3×10^{13}) human cells, with the number varying between 20 and 100 trillion depending on factors such as sex, age, and weight. The definition of terms include:

- Genomics. The genome is the complete sequence of DNA in a cell or organism.
- Epigenetics. How genes are turned off and on changes based on the environment and can be passed onto progeny.
- Transcriptomics. The transcriptome is the complete set of RNA transcripts from DNA in a cell or tissue. Although only 2% of the DNA is translated into protein, almost 80% of the genome is transcribed. This includes the coding RNA, short RNA, including microRNA, piwi RNA, and small nuclear RNA.
- Proteomics. The proteome is the complete set of proteins expressed by a cell, tissue, or organism.
- Metabolomics. Study of small molecules in the body (<1500Da), and there are more than 10,000 in the body.
- Lipidomic. Lipids or fats in the body.
- Microbiomics. The role of the microorganisms of a community. Microbes have been found in human skin, mucosal surfaces, and gut. The microbiome present in humans is very complex with at least 500

to 1000 species. The gut contains 39 trillion bacteria, slightly more than the 30 trillion human cells. [194]

We now understand that the answers to health cannot be found in any one omics. The microbiome influences gene and protein expression, which in turn influences the metabolome, and these processes crosstalk and regulate one another.

During the past twenty years, new modalities for precision therapies, CRISPR gene editing, RNA therapeutics and targeted protein degradation have proliferated. Innovative therapies powered by artificial intelligence (AI), CRISPR gene editing, and new sequencing technologies have increased returns on research and development (R&D), while enabling undruggable targets to become druggable. Increasingly, precision therapies are becoming multi-omic and curative, with mechanisms of action that span DNA, RNA, proteins, metabolites, and lipids.

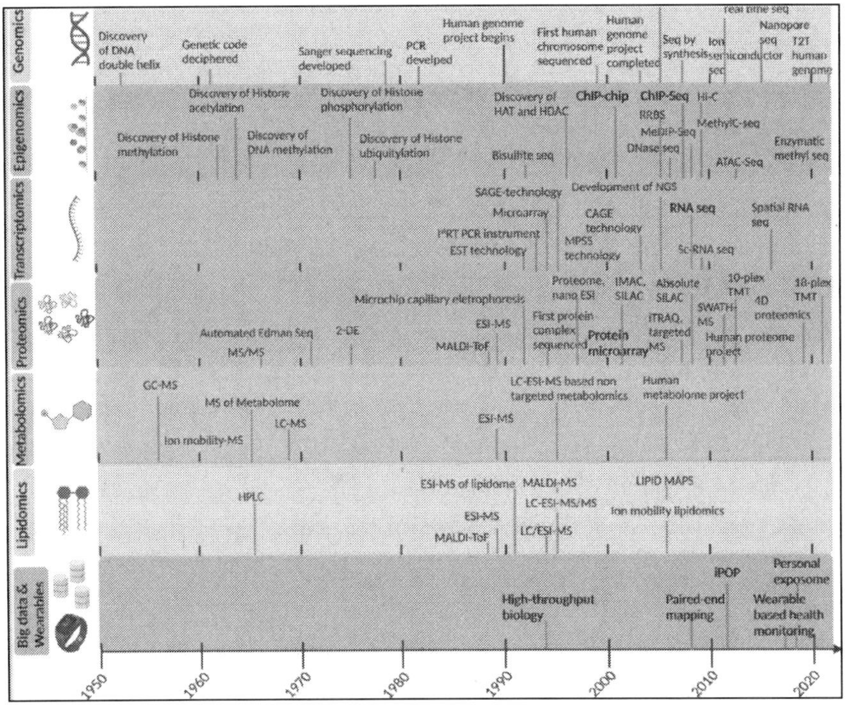

Figure 54 Timeline of major technological developments and milestones in different multi-omics analysis from PMC10220275. CC4.0

These relationships are explained in Figure 54 from Babu and Synder [193]. The diagram is too small to read here, so if you are interested look at the original. What we can see is for genomics, in 1950, Watson and Crick defined the DNA double helix structure. The Human Genome Project started in the 1990s and not completed until 2005. That has not stood still. In lipidomics, we had lipid maps by 2005; the Human Metabolomics project began around 2008, and 4D proteomics in 2019. CRISPR was developed in 1987, but it did not become a powerful tool until the early 2010s.

Multi-omic data feeds into novel Precision Therapies using emerging gene editing techniques that target and cure rare diseases and chronic conditions. Multi-omics is unlocking entirely new Programmable Biology capabilities, including the design and synthesis of novel biological constructs with applications across industries, particularly agriculture and food production. These advances would not have been possible without the disruptions in IT over the past 30 years, and it is difficult to comprehend or predict where we will be in another 20 years.

13.3Increasing Lifestyle Diseases

While exciting new technologies in health diagnosis and treatment and some measures are encouraging, it is not all good news. The first month of life is the most vulnerable period for child survival, with 2.3 million newborns dying in 2022. We examined in the population chapter that child mortality fell from 50% to less than 1% for most countries. Even over the past 25 years, neonatal deaths have fallen by 44%, and infant mortality rate globally has halved from 65 deaths per 1000 live births in 1990 to 29 deaths per 1000 live births in 2018. [195] Unfortunately, Sub-Saharan countries with a breakdown of society and famine still suffer from above-average infant mortality and we saw how average income inversely correlates with infant mortality. Experts estimate the Africa rate will fall to 25/1000 by 2050, the same rate as in Asia today. They have a way to go. The child in Iceland is 62 times less likely to die than in the Central African Republic. At the other end of life, life expectancy is rising in most countries, and the old are getting older. [196]

Welcome to old age. In the next few sections, we will further explore how some say we can all live to be over 120 years old. Most shudder at the thought of getting older because getting older means getting sicker! Can advances in health break that widely held view? Will society face up to higher expenditure on the health system? Will people eat better, or hope for advanced medical research translating into better health?

While other countries' longevity has increased, a wealth gap has opened up in the USA with decreasing life expectancy. [197] Across the lifespan, and across every demographic group, Americans die at younger ages than their counterparts in other wealthy nations. How could this happen? In a country that prides itself on scientific excellence and innovation, and spends an incredible amount of money on health care, the population keeps dying at younger and younger ages. A big part of the difference between life and death in the USA and its peer countries is people dying or being killed before age 50. The "*Shorter Lives*" report points explicitly to factors like teen pregnancy, drug overdoses, HIV, fatal car crashes, injuries, and violence. Older Americans have higher survival after age 75 than do their peers in other countries, and they have higher rates of cancer screening and survival, better control of blood pressure and cholesterol levels, lower stroke mortality, lower rates of current smoking, and higher average household income. But those achievements, it's clear, aren't enough to offset the other problems that befall many Americans at younger ages, namely opioids, guns and vehicle deaths. It is a salutatory lesson: don't take numbers at face value. Average numbers can hide some interesting data.

13.4 Obesity is Killing Us

The obesity patterns for different countries are the same slope - just a delay by 10 or 20 years.

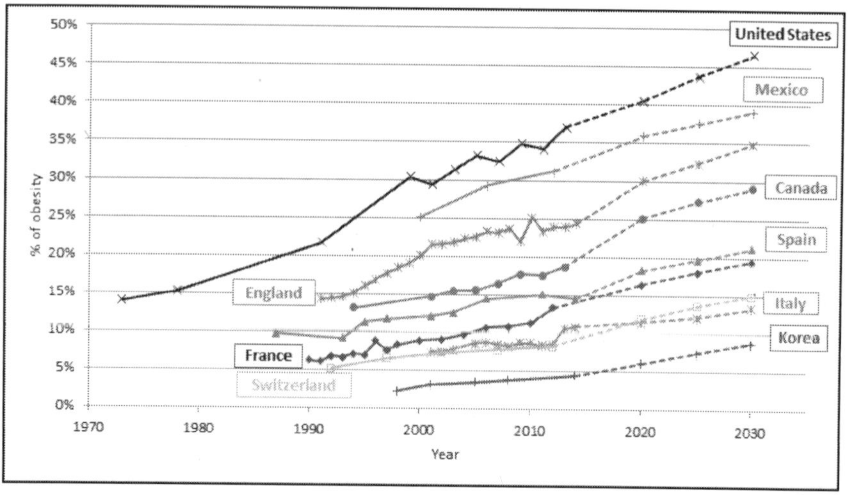

Figure 55: Obesity rates are rising in all countries.

Figure 55 does not include the overweight, which almost doubles the rates of obesity. We see the USA leading the pack but other countries such as England, Mexico and Canada chasing them. Associated with this rise in obesity and weight are poorer health outcomes and higher health burdens. [198] [199]. An increase in obesity is just one aspect of metabolic syndrome, as measured by increased girth, glucose intolerance, elevated blood pressure and increased weight. Metabolic syndrome is associated with higher cancer rates, higher heart disease and more musculoskeletal diseases. The increase in food products consumed correlates with this rise in obesity, but is this causation or correlation? We don't know. Other causes might include plastic and pollutants, lithium, or one of the multiple environmental changes. Prof Lustig says it is not due to gluttony and sloth but due to food choices. Changing food choices is a difficult challenge, as we saw in the food chapter, where just 12 corporations supply the majority of all food products, and over 60% no longer cook at home. It was always puzzling to me, as a New Zealander, why dinner with friends was always at a restaurant rather than at their homes, but the trend of not cooking at home explained that phenomenon.

13.5Cancer Rates

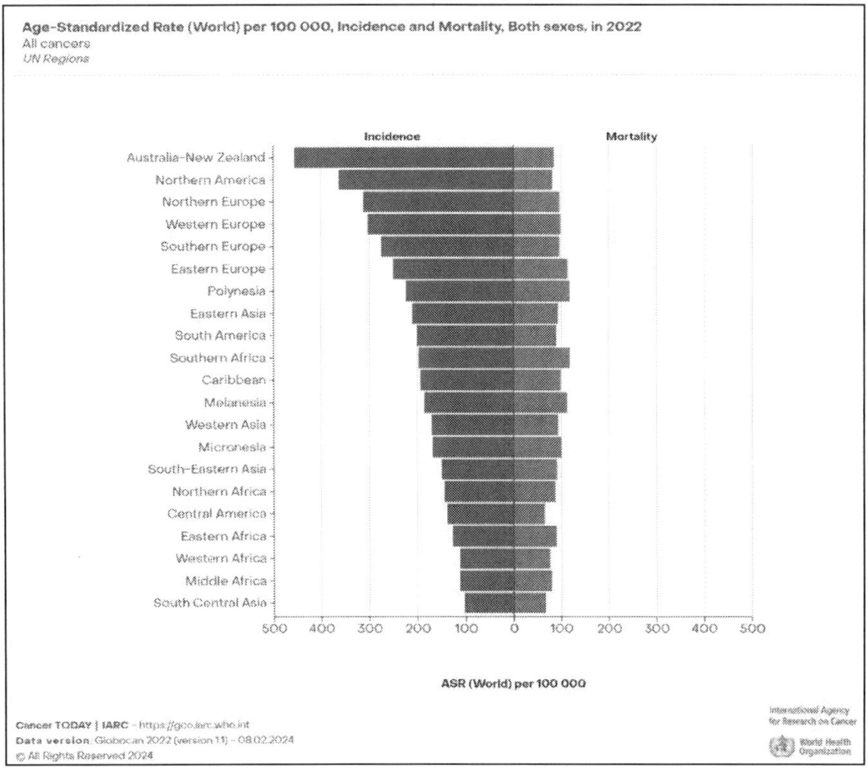

Figure 56 Cancer rates across the world from Cancer Today

Figure 56 shows that cancer rates are higher in Western countries, with the highest rate of cancer in Australia and New Zealand. A high level of skin cancer is the reason why these countries lead the cancer stakes. It has fallen with better diagnosis and world-leading cancer treatment programs. Cancer rates are falling in many countries. Overall, US mortality from all cancers combined declined by 26% from 1990 to 2015. The American Cancer Society has a 2035 goal of a 40% reduction from 2015. This reduction saved 1.3 million total deaths over that time and is expected to save 122,000 fewer per year by 2035. [200] Recent research indicates that US cancer mortality can still be significantly reduced by implementing known interventions more broadly and equitably.

Early detection of cancers generally increases longevity and improves treatment outcomes for many cancer types because the cancer is more localised. Removal or treatment is easier. E.g. more than 90% of people with

Stage I colorectal bowel cancer survive for 5 years or more, dropping to about 10% for the more advanced stages. High cancer rates in Australia and New Zealand are due to skin cancers, which are almost non-existent in many countries. Note that cancer mortality rates are generally similar across all countries. What stands out in the cancer data [201] is that mortality is about the same at about 80-100 per 100,000. What varies between countries is the incidence, with Australia and New Zealand high (skin cancer) followed closely by North America and Europe.

13.6 Cancer Types

The top 5 cancers of breast, lung, prostate, colorectal and cervical cancer account for more than half of cancers. Early detection decreases mortality. Detection delay means many die or have very painful recovery, simply because the cancer was caught too late.

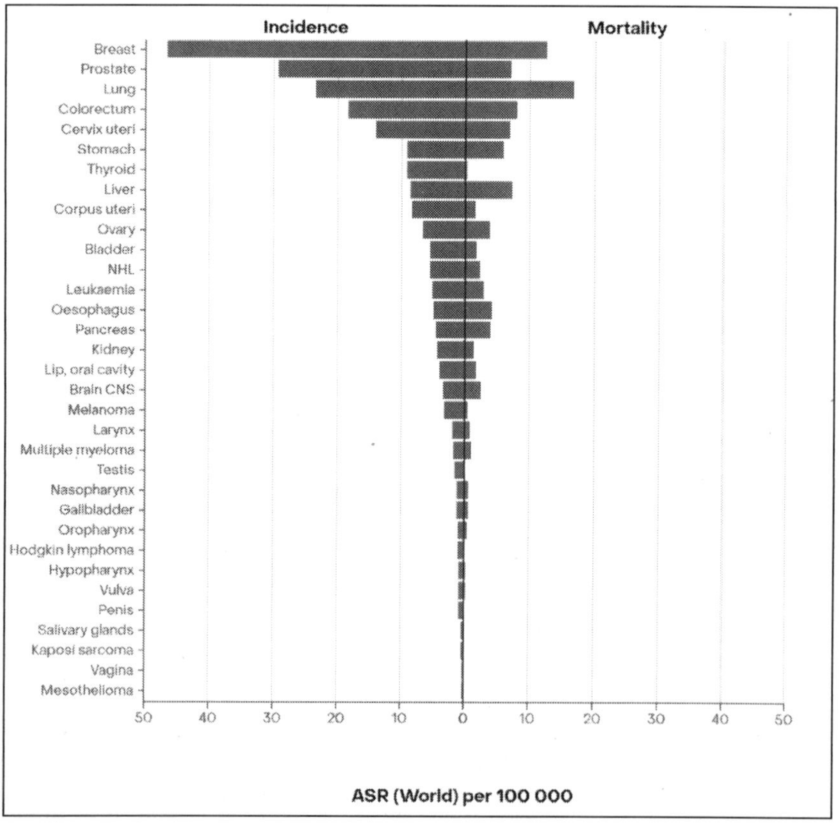

Figure 57 Age-Standardised Rate (World) per 100 000, Incidence and Mortality, Both sexes, in 2022 UN Regions from Cancer Today IRAC

Figure 57 shows breast cancer incidence the highest at over 40 per 100,000 or 4 times that of most other cancers. Ark Invest points out [202] that minimal residual disease (MRD) testing can detect cancer recurrence up to 20 months earlier than traditional imaging. The goal of cancer treatment is normally complete remission (CR). CR means they cannot find any cancer cells left in the bone marrow or organ. However the body always has some cancer cells – they just don't run amok. Clinical evidence shows ctDNA-based MRD testing can accurately detect molecular relapse 8–12 months before clinical relapse in early-stage breast cancer. Compared to advanced-stage breast cancer, detecting ctDNA in early-stage breast cancer is challenging and requires ultra-sensitive testing methods due to the low levels of ctDNA released into the bloodstream. Surgery and other treatments need longer follow-up. [203] Yet, more than eight million suitable cancer patients in the US are currently not reimbursed for MRD testing. Now, at only 10% penetration, MRD testing should become the standard of care for every cancer patient, with multiple tests per year and follow-ups for five years.

Along with MRD, multi-cancer screening is a revolutionary approach to early cancer detection. Unlike traditional methods, a single blood test can screen for multiple cancers simultaneously, streamlining screening and expanding its scope. The recent approval of Guardant Health's "Shield" colorectal cancer screening test suggests that the technology is proving its potential.

Advances roll out slowly. Australia has just initiated the rollout of free lung cancer low-dose CT scans for anyone between the ages of 50 and 70 who smoked and has no sign of cancer. Early detection improves lung cancer survival rates. How long were the trials? A decade. The screening will enhance early diagnosis and hence improve patient survival. The detection of Stage 1 cancers is expected to increase from 16% to 60%. Correspondingly, late-stage diagnoses are projected to decrease from 53% to 11%.[204]

A concerning report in 2025 from Australia [205] shows that cancer rates among younger Australians are nearly double what they were previously. Australians aged in their 30s and 40s are experiencing unprecedented and, in some cases, world-leading rates of at least 10 different types of cancer. Between

2000 and 2024, early-onset prostate cancer increased by 500 per cent, pancreatic cancer by 200 per cent, liver cancer by 150 per cent, uterine cancer by 138 per cent and kidney cancer by 85 per cent. Cancer is generally a disease of aging, so the cause of the increase is very concerning. Excluding the effect of better screening, something has caused the increase. Most experts believe toxins or toxic influences in the world around us are interacting with genes to cause malignant changes, and DNA markers are different to older cancer patients. Is it exposure to more chemicals? Is it the exposure to the plastics? Are these numbers worldwide or just limited to Australia?

13.7Heart Disease

Global deaths from heart and circulatory diseases average 56,000 people each day, or one death every 1.5 seconds. They are the world's biggest killers. Globally, heart and circulatory diseases killed an estimated 9.8 million men and 9.2 million women in 2019. Heart disease accounts for over 1 in 4 (27 per cent) of all global deaths. [2]. While the total number of deaths is increasing, the age-standardised numbers have been decreasing. Heart disease in women is rising.[206] Heart disease rates differ widely.

- High-income countries: Heart disease death rates have generally declined over the past decades due to better prevention, early detection, and treatment.
- Low and middle-income countries: Heart disease deaths are now rising, often outpacing declines in other causes of death. This is likely due to several factors, including:
    - Increased risk factors like smoking, unhealthy diet, and physical inactivity.
    - Limited access to healthcare services for prevention, diagnosis, and treatment.

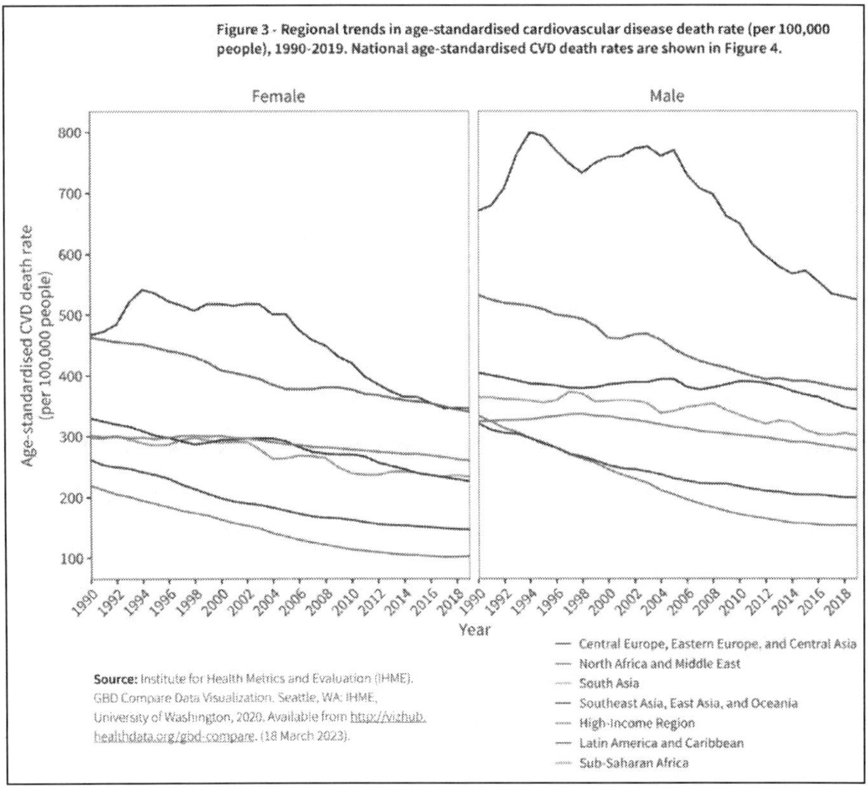

Figure 58 Cardiovascular death rates from World Heart Federation

Figure 58 from the World Heart Organisation shows that cardiovascular deaths have been falling across all countries by approximately 25% between 1990 and 2019, and the reason is due to improved management. Heart disease was always considered the silent killer, with the first indication the last. Education, along with prevention, has been the primary factor.

13.8 Aging and Dementia

Aging is a complex process and is often associated with disease. Since the Greeks, humans have tried to understand aging with little success. Multi-omics is providing some tools to understand what is going on, and a recent study by Shen et al [207] "Revealed consistent nonlinear patterns in molecular markers of aging, with substantial dysregulation occurring at two major periods, at approximately 44 years and 60 years of chronological age. Distinct molecules and functional pathways associated with these periods were also identified, such as immune regulation and carbohydrate metabolism that shifted during the 60-year transition and cardiovascular disease, lipid and alcohol metabolism changes at the 40-year transition." Anyone reading this who is over 40 or over 60 has probably experienced those changes. Often, these periods of change coincide with other physical changes, including females going through menopause, men losing libido or developing that middle-aged spread, and a shift in focus in life.

13.8.1 Lifestyle is 84% of Life Expectancy

What we currently realise is life expectancy is incredibly complex. We also know that less than 16% can be explained by genetics. Knowing which lifestyle behaviour will impact life expectancy is a challenge? Some factors are known, but many are unknown. It was only a few years ago that researchers realised that epigenetics was real and that external factors could change gene expression. Research is slowly knocking off the unknown causes, and the microbiome is an integral part of the equation. Microbiota are more important in keeping us well than in making us sick.

13.8.2 How to Live to 200 Years, or 120

Some medical researchers express confidence that it is possible to modify lifestyle to live to 120 years or even longer, and there is no valid reason why 200 years is not possible. You may not want to, or some factors that reduce longevity are known. E.g. 9% of Americans are diabetic, and 33% are pre-diabetic. Those numbers are rising, and higher glucose levels reduce life expectancy. Lifestyle changes and behaviours can be influenced by technology and feedback practices, including continuous glucose monitors and real-time feedback.

Dr Snyder, director of the Stanford Centre for Genomics and Personalised Medicine [168], [169], says there is no physical reason why people cannot live to 200 years. To do so, one would want to live a pain-free and productive life, die quickly at the end, and not linger for decades. Are we confident that research will advance to the point where we know what a healthy baseline is and how to live? Maybe.

- Genetics. Genetics only contributes about 18% to longevity.

- Exercise, and especially weight lifting. Weights reduce muscle mass loss (sarcopenia). Muscle mass decreases approximately 3–8% per decade after the age of 30 and this rate of decline is even higher after the age of 60. I.e. over 0.2kg to 0.6 kg per year (2 to 8 kg per decade). The simple message is that exercise buys a further 5 years of life. Five years is a good start.
- Miracle drugs. Meanwhile, scientists are looking for that miracle drug. One drug, an antibody that attacks interleukin-11, has increased the lifespans of laboratory animals by 20% to 25%. [208], [209] That antibody is now in clinical trials with humans with lung fibrosis. Other type 2 diabetes drugs, including metformin and rapamycin, are also in active trials.

Drugs are easier for people than calorie restriction, which has been shown to extend life in other animals. It seems unthinkable to treat every 50-year-old for the rest of their life, but many do it now.

13.8.3 We Take Pills, Lots of them

We often live with friends and family who have health issues. The number of pills that adults take is rising. The USA numbers are astonishing but most Western societies are on this treadmill.

- 90% of 65 years and older are on prescription medicine.
- 75% of 50-64-year-olds.
- 50% of 30-49-year-olds.
- 39% of 18-29-year-olds.

Australia is not far behind; two-thirds of people aged 75 years and over take five or more medicines, including over-the-counter and complementary medications. Some would argue this is good as illness is being treated, but most “illness” is due to poor lifestyle. Metabolic syndrome disease – elevated blood pressure, high cholesterol, increased girth, insulin resistance and type 2 diabetes are the reasons people are on medication.

13.8.4 Dementia Crisis

One of the significant changes as people age is the risk of dementia. We probably know someone with dementia or are concerned for ourselves. None of us wishes to lose our faculties. [210] The hit list for dementia is:

- Dementia is ranked seventh among the leading causes of death in the world. [211]
- 1.5 million deaths a year
- The types of dementia are Alzheimer's (60- 70%), Lewy Body (5- 10%), vascular (5- 10%), and frontotemporal (5- 10%).
- More than 60 million people are living with dementia.
- Annual cost to the global healthcare economy of around US $1.3 trillion (almost £1 trillion).[212]
- Projected increases are primarily due to more older people. People are living longer.
- Cases are expected to triple to 153 million new cases by 2050.
- The lifetime risk of dementia in the USA after age 55 years is 42%.
- USA new cases double from 514,000 to a million by 2060, driven by higher rates in the underreported black population. [213] [214]
- Black adults and APOE ε4 carriers have higher lifetime risks ranging from approximately 45% to 60% in these populations.
- Any previous decrease in dementia rates is a function of better management of cardiovascular disease.
- More women than men get dementia. The female-to-male ratio of 1.69 is not explained by women living 6 years longer than men. At 95 years old, women have a 48% probability of dementia, and men have a 35% probability of dementia.
- The top 3 risk factors are high body mass index, high fasting plasma glucose, and smoking.
- 45% of dementia cases could potentially be prevented by reducing exposure to 14 modifiable risk factors common throughout the world.

Given the prevalence, the impact on family and health and the preventable nature we might well ask why more is not being done? Similar to metabolic syndrome as a consequence of obesity, most would rather take a pill than deal with the root causes. Dementia is not just a lifestyle disease. It also has a huge impact on family, the aged care industry and health services.

13.8.5 Exercise for Prevention of Dementia

At this stage there are no magic bullets. The research is compelling that the most effective treatment is exercise. Until we know the cause of dementia, exercise! The probability of dementia is reduced by 30% just by walking. Yet many do not. How to get them to do more physical exercise?

In a landmark study in Australia [215] internet-delivered lifestyle intervention resulted in significantly better cognition in older adults. The 4 interventions include a Mediterranean diet, 5 hours of exercise per week, 3x45 minutes internet-based cognitive exercises and mental health interventions. The ability to shift focus and memory, with more exercise, is remarkable. One researcher in the USA, Prof Wendy Suzuki, has numerous publications on techniques to change behaviour. [216] Another study shows that simple tactics, such as using skin lotion twice a day for 30 days, can reduce skin inflammation. Weirdly, that improved epidermal function, reduced circulating cytokine levels and trial participants gained cognitive function!

VO_2 max, or maximal oxygen consumption, is a measure of your body's ability to use oxygen during intense exercise and is a key indicator of health and lifespan. VO_2 is likely one of the primary factors worth focusing on for longevity. Unfortunately, building up a good VO_2 max takes a little bit longer than getting strong – but there are things you can do to speed the whole process along. One is the so-called "Norwegian protocol": working out at a high intensity for four minutes, then at a lower intensity for three minutes, then repeating – four times altogether. With a warmup on top, you're done in a little more than 30 minutes, and if you do that once a week, any other running workouts can be more ponderous or plodding.[217] Programs such as HIT (high intensity training) used to be the exercise of choice. HIT is short bursts of high intensity, rest, and repeat. That has its own issues, particularly if the person had not built up a level of fitness prior. Workouts can also be very simple with no need for dozens of movements. All you need to focus on is what many trainers call the fundamental human movements: push, pull, squat, "hinge", and carry.

In Figure 59, the "S" diagram [218] shows 13 modifiable risk factors of dementia and the numbers are the percentage reduction in cases of dementia if that risk factor is eliminated. Over 45% are potentially modifiable.

- In early life, education is critical and accounts for 5%.
- Midlife factors have their roots in childhood or even at birth. Some of these midlife risks need to be addressed in childhood or the teenage years. For example, 80% of adolescents living with obesity will remain this way when they are adults. We also know now that birth weights of children born today are over 200 g higher than those of children born 20 years ago. Those bigger children turn into oversized teenagers who turn into obese middle-aged people. The same applies to high blood pressure and lack of exercise. Similarly, virtually all adults who smoke or drink will have started these unhealthy habits in or around adolescence. Altering health behaviour that has already been established is notoriously tricky. And second, most high-risk individuals targeted in middle age will almost certainly have been exposed to the damaging effects of these risk factors for many decades already. Recently high LDL cholesterol has been added to the risk factors – at over 7%.
- Late life – 13% is still modifiable and include social cohesion, air pollution and vision loss.

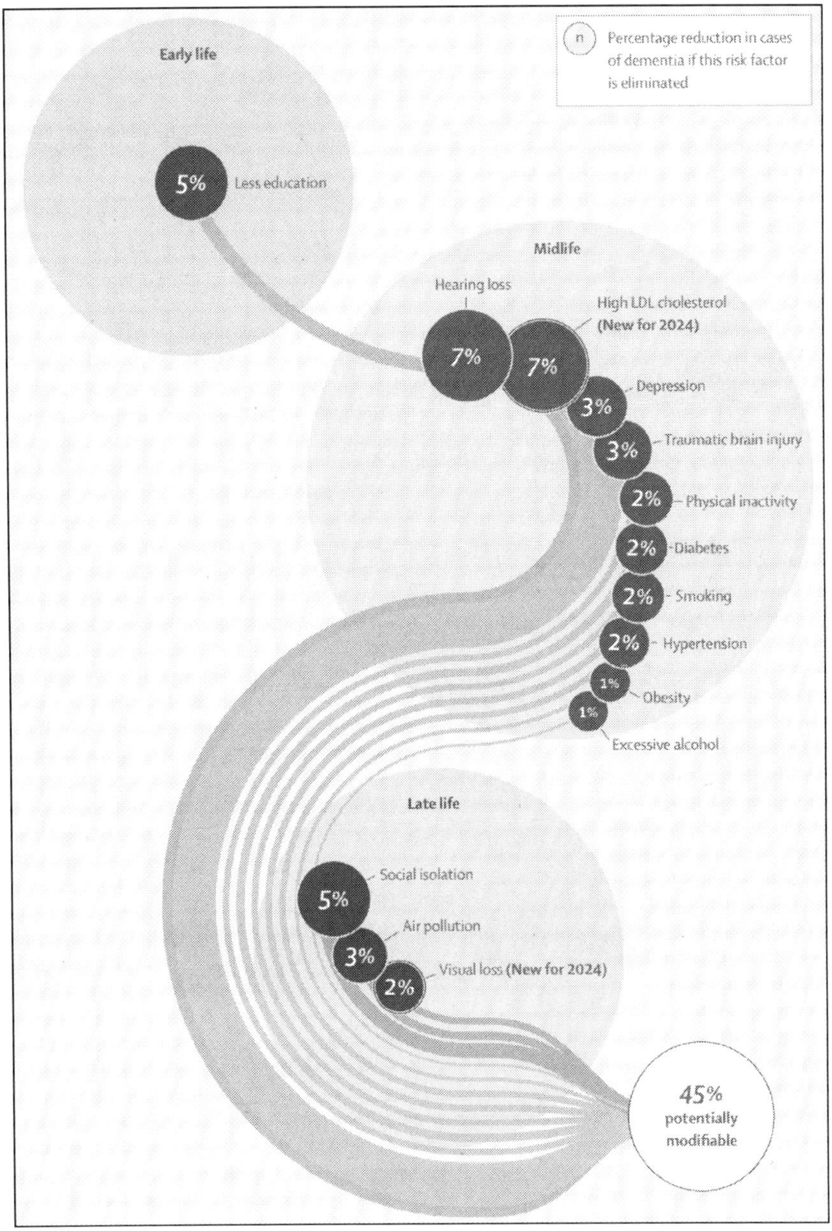

Figure 59 Population attributable fraction of potentially modifiable risk factors for dementia

Increasing evidence says many of the differences in brain structure and function associated with dementia in older adults may have at least partly existed since childhood.[219]

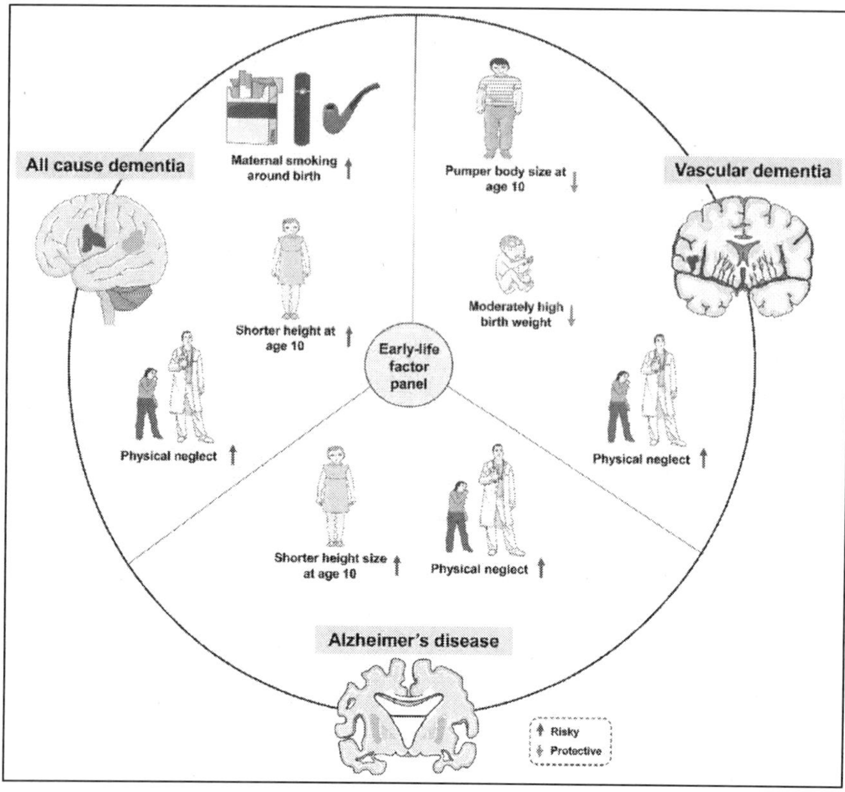

Figure 60 Early childhood increases all cause dementia risk

In Figure 60, Ya-nan Ou demonstrated physical neglect and high birth weight increase the risk of dementia and broke it down into all cause dementia (maternal smoking, physical neglect) , vascular dementia (high birth weight). Being shorter than expected at age 10 was a key marker as was physical neglect.

13.8.6 Dementia Diagnosis

Meanwhile, advances in AI and measurement technology are improving early diagnosis, which is critical, as early treatment is most effective in slowing down the disease.[220] Using AI plus cognitive tests and MRI scans could correctly identify 80% of those who developed, as well as those who stabilised. One study demonstrated a 90% accuracy in the early diagnosis of dementia through blood tests [216], compared with 61% for general doctors and 73% for specialist dementia doctors. Diagnosis is one thing, but what treatment works?

13.8.7 Dementia Treatment

Treatment for dementia is a more complex issue. Donanemab (Lily's Kisuna) is a monthly injection and can be used for up to 18 months. While it is FDA, UK and TGA-registered [158], at $48,000 per annum, it has a long way to go before becoming affordable.

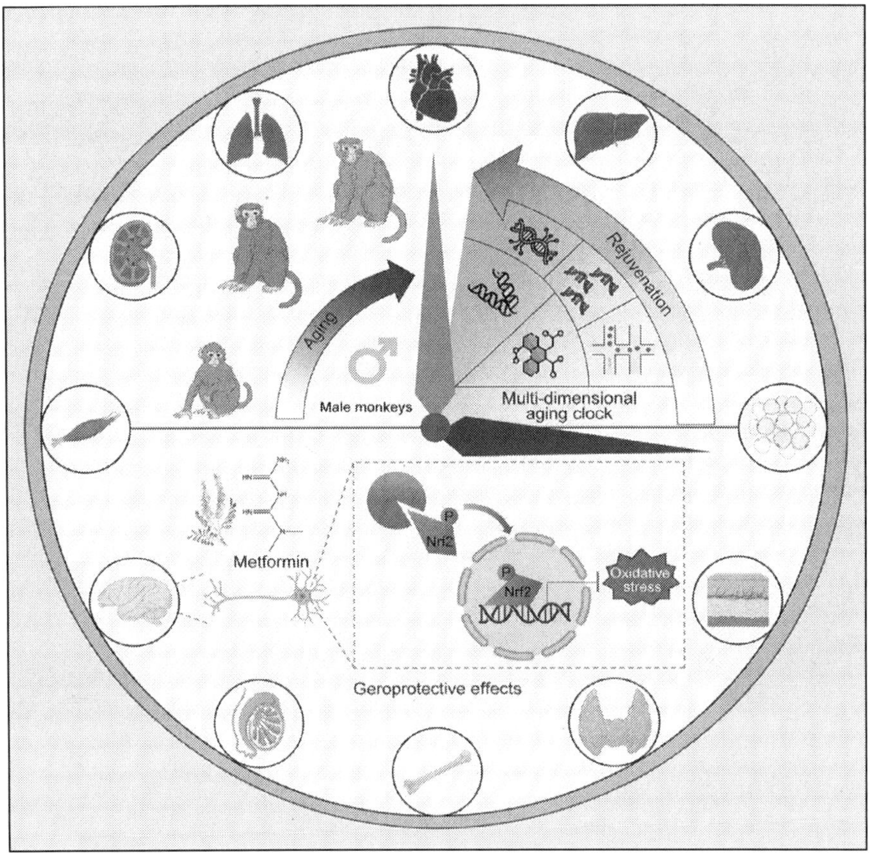

Figure 61 Geroprotective effects from metformin in aging male monkeys and rejuvenation of tissues.

Figure 61 shows that metformin in monkeys reduces cognitive decline by up to 6 years, which translates to 18 years in humans. [159] Metformin was first described in 1922, has been in use as a medication since 1957, and received FDA approval in 1994. It has also been used for a range of diseases, including diabetes. Will this latest publication lead to a spike in consumption? The gero-protective effects of metformin include reducing damage to the heart, lungs, kidneys, and muscle tissue. Ozempic and other GLP-1 antioxidants have similar actions, and they too appear to reduce "inflammation". Other monoclonal antibodies, such as lecanemab and aducanumab, are also at the forefront of new products aimed at slowing down or reducing tau and amyloid plaques. While they are not silver bullets, they represent a path toward developing more innovative therapies that target the underlying biology of dementia.

How effective will these treatments be? $48,000 pays for a lot of personal fitness trainers!

13.9 Parkinson's Disease

Parkinson's disease (PD) is a progressive neurodegenerative disorder that affects 2%–3% of the global population over the age of 65 years and is increasing globally. PD affects more than six million people globally and is projected to triple by 2040. Celebrities such as Michael J Fox are the face of PD research. We will explore it further as an example of a disease once thought untreatable, but omic studies combined with AI [221] is leading to new understanding and a raft of promising treatments. Using machine learning, the researchers were able to speed up the initial screening process tenfold and reduce the cost by a thousandfold. [222]

13.9.1 Factors Contributing to PD rise

- **Aging Population**: The most significant risk factor for PD is age. As the global population ages, the number of people susceptible to PD increases. PD rises sharply over 60 years of age.
- **Environmental Factors**: Exposure to certain toxins, pesticides, and industrial chemicals is thought to increase the risk of developing PD. These include:
    - Paraquat was widely used 4 decades ago, with a risk of up to 150%.
    - Rotenone, an insecticide, has been shown to cause PD in animal studies.
    - Organochlorines, including beta-hexachlorocyclohexane.
    - Industrial chemicals such as trichloroethylene. TCE is used in various industrial processes, including degreasing and dry cleaning. TCE has been associated with a 500% increased risk

of PD. It's a widespread contaminant in groundwater and can even vaporise into homes and workplaces.

  - Perchloroethylene (PCE) is used in dry cleaning and other industrial applications.

- **Lifestyle Factors**: While less definitive, studies suggest that lifestyle factors such as lack of exercise, poor diet, and head injuries may play a role in the development of PD.
- **Genetic Predisposition**: Family history of PD increases the risk of developing the disease. Although genetic factors play a lesser role compared to aging and environmental factors, they still contribute to the overall rise.

13.9.2 Recent Advances in Understanding PD

Significant progress has been made in understanding PD, leading to potential new treatments and preventative strategies with different approaches:

- **Genetics:** Specific genes are linked to PD. They give insights into underlying mechanisms of the disease and potential targets for gene therapy.
- **Lewy Bodies and Alpha-Synuclein**: Research on Lewy bodies (abnormal protein clumps in the brain) and alpha-synuclein (a protein within Lewy bodies) is helping to understand the disease process and develop therapies to slow or stop its progression.
- **Neuroinflammation**: Inflammation in the brain plays a role in PD. This has opened new avenues for potential treatments targeting the immune system.
- **Gut-Brain Axis**: The connection between the gut microbiome and the brain (gut-brain axis) suggests a potential role for gut health in PD development. Maybe this could lead to dietary or microbiome-based interventions.
- **Early Detection and Biomarkers**: The hunt is on for biomarkers that can help detect PD in its early stages, even before symptoms appear. This would allow for earlier intervention and potentially slow the disease's progression.

While there is still no cure for Parkinson's, recent advances offer hope for better treatments and potential preventative strategies in the future.

- Disease-Modifying Therapies aim to halt or slow progression.
- Alpha-synuclein targeted therapies to reduce its production, aggregation, or promote its clearance. These include antibodies against alpha-synuclein and small-molecule inhibitor drugs to block the formation of alpha-synuclein aggregates.
- Gene silencing therapies to reduce the production of alpha-synuclein.
- Neuroprotective therapies to protect neurons from damage and degeneration including GDNF (glial cell line-derived neurotrophic factor, NPT200-11 which improves mitochondrial function., which is impaired in PD and anti-inflammatory therapies.
- Symptomatic management of symptoms of PD have been around for a while including levodopa which is the gold standard for treating motor symptoms, dopamine agonist, MAO-B inhibitors, COMT inhibitors and deep brain stimulation.

13.9.3 How AI Is Helping PD Research

PD is a good example of how AI is assisting research. This rise in AI will lead to new treatments, including cell-based therapies and gene therapy. Horne et al. show in Figure 62 how machine learning has fast-tracked the identification and treatment pathway. In the docking strategy, they checked over 2 million compounds for docking. They identified and tested 68 and had 4 hits on potential drug solutions.

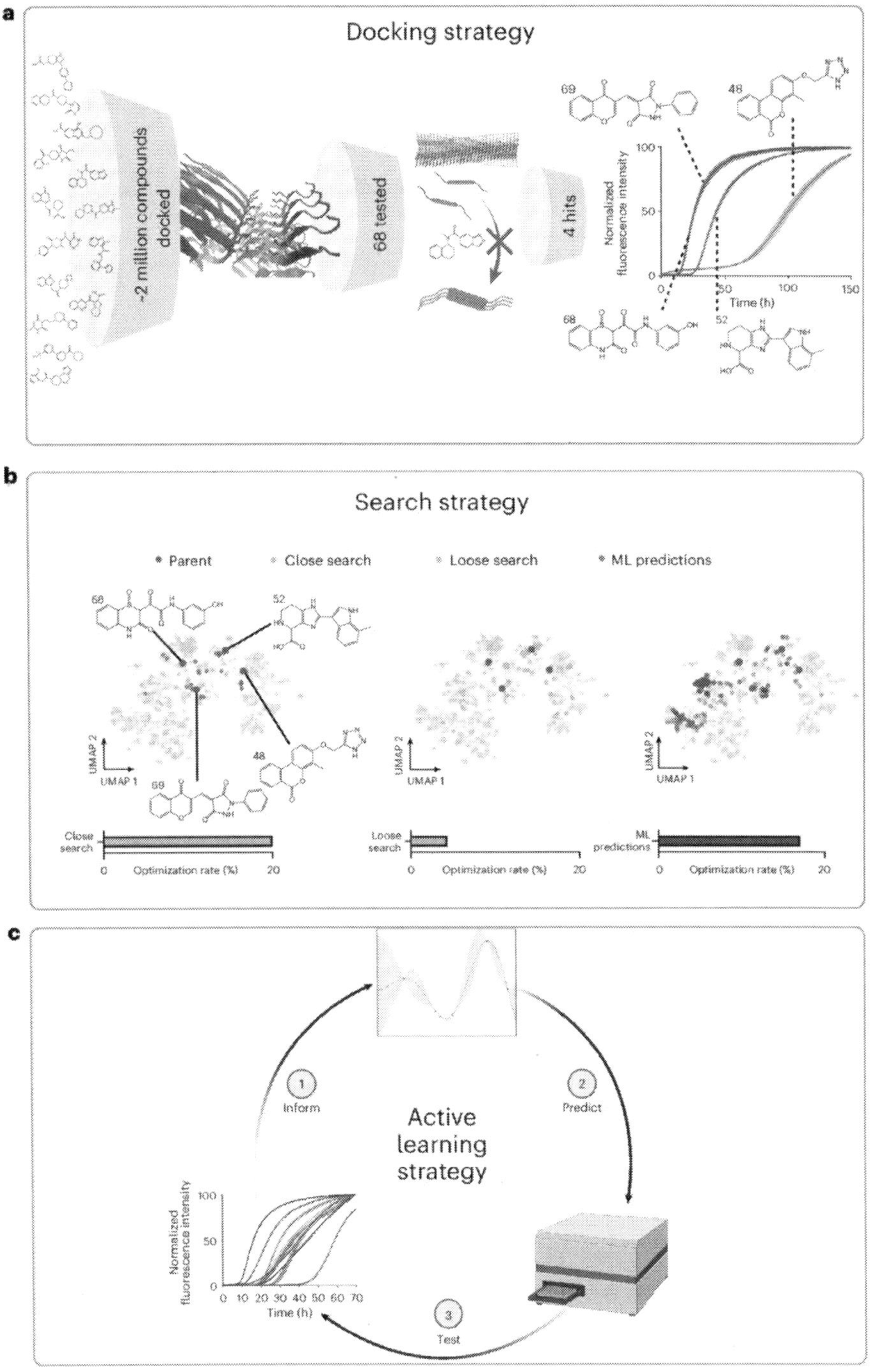

Figure 62 How AI accelerates the finding of drugs for Parkinson's disease from Horne et al 2024

13.10 Huntington's Disease

Huntington's Disease (HD) is a good example of a disease that has a very clear genetic basis. There is currently no cure, and it provides a case study on whether people get tested if they are at risk. HD is an inherited, fatal neurodegenerative disorder caused by an expanded CAG repeat in the HTT gene, producing a toxic huntingtin protein and progressive loss of motor, cognitive, and psychiatric function.

The condition was described long before it had a name. In 1860, a Norwegian physician, Johan Christian Lund, reported a high prevalence of dementia with jerky movements in the remote Setesdal region of Norway. The disease's familiar name comes from George Huntington, an American physician from Long Island who wrote "On Chorea" in 1872 at age 22, a year after his medical degree – making him the describer, not the "first person to get it." HD itself almost certainly stretches back centuries; close to 1,000 descendants of original colonial-era carriers are known to have had the disease over three centuries, with some unlucky sufferers tried for witchcraft. The most striking case for "descendants" is Venezuela. The largest known cluster is in the Maracaibo area, home to roughly 14,000 affected people, traced back to a North European sailor who landed in a fishing village in Zulia state around 1800. This Venezuelan population became the focus of US research expeditions from 1979 onward, leading to the discovery of the gene's chromosomal location in 1983 and the gene itself in 1993. Iceland, by contrast, has very low HD prevalence, with nearly all cases tracing to one couple living in the early 19th century.

13.10.1 Current breakthrough research

There is no cure currently. There is a potential gene therapy, AMT-130, from uniQure. It works by halting production of the toxic huntingtin protein, targeting the messenger RNA rather than altering DNA – a key safety advantage. After three years, the high-dose group showed 75% less disease progression than an untreated comparison group, also with a significant drop in a neurodegeneration biomarker. The therapy is delivered directly to the brain via an 8 to 20-hour surgical procedure.

UniQure expects to submit data to the FDA and seek approval in early 2026, with a likely price comparable to other gene therapies (around $2 million or more). AMT-130 has already received breakthrough therapy, orphan drug, and fast-track designations, and it's the first AAV gene therapy for Huntington's to receive the European Medicines Agency's Orphan Medicinal Product Designation. [223]Success may pave the way for similar approaches to other neurodegenerative diseases.

This is the first time in HD's history that a treatment has been shown to slow — rather than just manage symptoms of the disease itself.

13.10.2 Do You Want Testing?

Dr Nancy Wexler was a key researcher who spent her life identifying the gene, and she herself had HD. She writes [224] *I realized I hadn't thought through all the ramifications, both psychological and social, of what a positive test might mean in my life. I wasn't sure I could live with the foreknowledge that one day the dreaded symptoms would emerge. I preferred to meet the devil when he was on my doorstep.*

HD highlights challenges for gene therapy. Success may pave the way for similar approaches to other neurodegenerative diseases. Can you afford a $2 million treatment?

13.11 Chemicals Affect Health

No one knows exactly how much chemical pollution is a crisis in the making. How did 200 to 700 chemicals get into our bodies? There are more than 150,000 different man-made chemicals, with 2,000 new chemicals added every year.

How do they actually affect us? The $7.8 trillion industry continues to develop new chemicals. We know they are not supposed to be there. [225] The chemical industry is one of the largest manufacturing sectors and BASF, Sinopec and Dow are the 3 biggest. Producing over 2.5 billion tonnes of commercial chemicals annually, the industry serves as the invisible backbone of modern civilisation and acts as the largest single purchaser of its own products. Some of these, such as sulphuric acid, ammonia or Agent Orange we may have heard about. The majority are man-made chemicals and have made their way into every corner of Earth and human life - to the extent that industrial toxins are routinely found in newborn babies and their mothers' breast milk. The snow at the peak of Mount Everest is so polluted that it doesn't meet drinking water standards.

Lead, arsenic and 14 other chemicals are found in most tampons globally, yet we know there is no safe level for lead. [226] As New Atlas reported, a major concern of researchers was that lead was in all tampons. Not one was free of lead, and as pointed out there is no safe exposure level. Any lead that leaches out of a tampon enters the bloodstream and harms health. Lead is stored in the bones, replacing calcium, and can stay there for decades. .

The chemical industry continues to grow rapidly - even though exposure to hazardous chemicals is linked to an increasing global burden of disease. The industry is also a major CO_2 polluter emitting about 2 billion metric tons of CO2 ($GtCO_2$) per year, accounting for about 5% of global GHG emissions. Examples of the health effects linked to exposure to some of the hazardous chemicals:

- Cancer.
- Reproductive system disorders, including endocrine disruption, male infertility, earlier onset of puberty in girls, impaired immunity and obesity.
- Nervous system disorders, including cognitive impairment.

We worry about the "forever" chemicals, but we don't know what impact these chemicals have on us. There is a serious gap in our understanding. Many suspect [227] the oestrogen-like chemicals affect our health in ways we simply don't yet comprehend.

There is good reference material on chemicals and health. Forty-five years ago, as a junior weed scientist, I tended to be blasé about the impact of chemicals on society and on human health. Knowing and seeing the direct impacts of chemicals on health today, less blasé. My focus has been firmly on CO2 emissions and the existential threat to our way of life from burning fossil fuels, but the impacts of chemicals are certainly real for some, if not all. Where there is smoke, there is fire, and over the next decade, our improved measurement techniques will undoubtedly reveal the catastrophic impacts of unrestrained chemicals.

13.12 Plastics Effects Health

We are living in the "Plasticene" era. An era in which plastics have permeated every aspect of our lives like an epidemic. We now use over 60 kg per person per year, up six times over the past 20 years. Microplastics are impossible to avoid. Plastics production, which began less than a century ago, reaches an all-time high each year, and the material takes hundreds, if not thousands, of years to degrade, creating trillions of microplastics in the process. In 1950, the world produced just two million tonnes. It now produces over 450 million tonnes, twice as much as in 2000.[228] Only 9% of this plastic is recycled, with the rest ending up in landfills, incinerators, or polluting the environment. Plastic production is set to double by 2040. The Global Plastic Pollution Treaty, a United Nations-led international legal agreement, aims to end plastic pollution, but the 175 nations negotiating the new treaty, which began in 2022, were stuck at the 5th meeting in Busan, South Korea. [229] Major producers, including China, Russia, Iran, Saudi Arabia, and the United States, aim to increase production. Others push for stricter global control of plastic production.

Recent headlines such as "microplastics are infiltrating brain tissue, and there's nowhere left untouched" highlight the global pollution from plastic. Researchers looked at 12 brain samples from people who had died with dementia, including Alzheimer's disease. These brains contained up to 10 times as much plastic by weight as healthy samples, at about 0.5% plastic by weight. [230]

The average brain has up to 10g of plastics, [231] *about the weight of a crayon or 2 credit cards*

Plastics are gobbled up by immune cells, which then pile up and block small blood vessels in the brain. [232]

Newton lived for 84 years in the late 1700's, and his succumbing was probably due to his mercury levels – some 40 times the average. However, he was an avid alchemist who frequently experimented with, tasted, and boiled mercury in his attempts to convert base metals into gold.

Many other studies have detected tiny shards and specks of plastics in human lungs, placentas, reproductive organs, livers, kidneys, knee and elbow joints, blood vessels and bone marrow. On average, microplastic levels were about 50% higher in brain samples from 2024 than in those from 2016. And brain samples contained up to 30 times more microplastics than samples from a person's liver and kidneys. But it is not just plastic, but a whole raft of other chemicals associated with plastic including plasticisers, phthalates and other chemicals. Almost all plastics are derived from fossil fuels and have been found to contain over 16,000 chemicals, many of which are considered hazardous. Over 6,300 have high exposure potential, including more than 1,500 that are known to leach from plastics. [233]

Shockingly, despite evidence of harm, the US regulates only 6% of the chemicals used to produce plastic, leaving potential health risks from most of the remaining chemicals unchecked. A staggering 66% of plastic-associated chemicals lack available hazard data. Of those that do, more than 4,200 (around 75%) are already known to be hazardous to human and/or environmental health. For example, phthalates and bisphenol A are known endocrine disruptors and have been associated with multiple health impacts at current levels of exposure in the general population, including type 2 diabetes in adults and developmental changes in newborns. It is not just pets [238]; data show that crawling on the floor puts babies at increased risk of inhaling microplastics in household dust. And since babies chew on everything, they are also more likely to ingest microplastics. [234] This realisation that plastics, promoted for food safety, to avoid spoilage and reduce costs, might well turn out to create a health disaster. Tracey is unaware of the scale of the disaster unfolding. She is awash with chemicals in her house, bathroom, food and drink.

13.13 Is Measuring Key To Health?

We used to study one gene at a time and it was cumbersome. Now, technology allows the study of all the genes at once. This system's approach, bringing genetics to medicine, has led to more understanding and powerful diagnoses. Cross-referencing an individual's genetic sequence – their genome – against known elements of "Big Data" brings new insights. Elements of genomics are already being incorporated on a widespread basis, including prenatal disease screening and targeted cancer treatments. While genetics accounts for only 16-

18% of one's lifespan, it does provide a blueprint for physical appearance, disease propensity, and behaviour. Health is not genes per se, but the regulation of them. These flow from gene regulation/epigenetics to the manufacture of proteins, lipids and other metabolites and all this together with interaction with microbiota. Turning genes off or on is now the new medicine. A few genes predict mutations for some diseases like breast cancer genes, but these are rare.

AI is essential in understanding the whole system. An example is ALS (Amyotrophic Lateral Sclerosis), which is a nerve/spinal disease. Traditional genomics suggested that seven genes were the cause, but with the aid of AI, researchers found over 695 genes that control ALS expression. These advances would be impossible to resolve without big data and AI.

13.13.1 Even More Measurements

Only 15 medical tests are common practice today, and it turns out they don't affect health. In contrast, research has demonstrated that there are hundreds of measures that can be done with modern DNA profiling, including protein molecules, RNA, body and gut (microbiome). A Stanford project 11 years ago monitored 109 people and took over 1,000 measurements from each. [235] The authors said "*We discovered more than 67 clinically actionable health discoveries and identified multiple molecular pathways associated with metabolic, cardiovascular and oncologic pathophysiology.*" Which means, in plain language, the measures caught diseases before they had any symptoms. Within the first 3 years, the study caught 67 people with presymptomatic diseases, including one with lymphoma, two people with precancerous conditions, pre-ovarian and pancreatic cancers, and two people with severe heart issues. In-depth data analysis is necessary to comprehend over 800 measures, including whole-body MRI. If cancer is caught early, medicine can treat the cancer. Catching these diseases after symptoms appear is disastrous and likely to fail.

The Lancet reports [236] one in five people is at high risk for atherosclerotic cardiovascular disease and aortic valve stenosis due to high lipoprotein(a), yet a simple test, done once in a lifetime, can identify if they fall into this high-risk group. Lipoprotein(a) lowering drugs are not yet available, although there are some in development. How many have had a test? Almost none.

A key part of the measurement is using wearables to monitor and adjust lifestyle. Snyder says wearables are our future for monitoring health. [237] [238] Over 50% of the global population owns a smartphone. Combined with a smart watch / monitor it can provide most of the monitoring. Wearables are incredible health monitors, and wearing them 24 hours a day can measure heart rate variability – a key indicator for many diseases. Blood oxygen measurement by a smartwatch is not very accurate, but it is good enough to detect changes. Delta maps are more important than absolute values. Snyder even detected Lyme disease using a wearable before he had symptoms. Unlike conventional methods – which you likely use only at doctor visits, maybe once or twice a year – wearables can be worn anywhere, anytime, continuously tracking data and revealing health trends, helping doctors "fill in the gaps" when making a diagnosis or prognosis. Medicine is undergoing a radical transformation to move to point-of-care, away from centralised medicine as costs continue to rise, and to medicine at home, enabled by remote diagnosis.

13.13.2 Is Measuring Effective

Other measurements by smartwatches include dry skin, which is an indication of diabetes, blood pressure and haemoglobin levels. Continuous blood glucose monitors that measure interstitial glucose provide a baseline, and AI can then predict ongoing glucose levels. These wearables can detect Covid with an 80% success rate and 3 days before symptoms, which is as good or even better than rapid antigen tests. Moreover, stress is picked up, or even mental health stress. Privacy is at the forefront of this sensitive data and has cybersecurity and privacy risks.

The approval for sale by the FDA in early 2024 of the over-the-counter continuous glucose monitors (CGM) made by Dexcom was meant for Type 2 diabetics who don't use insulin but want to avoid regular finger-prick tests. And some new CGM start-ups, like Signos, Nutrisense and Levels Health, now market prescription devices off-label as tools for energy, mood and metabolism.[239] Expert researchers and scientists argue that the evidence for CGMs improving the health of non-diabetics is effectively non-existent. They insist the devices are, at best, a distraction and, at worst, could lead to dangerous fixations. Contrast these experts with Snyder who say wearables and AI will catch disease before symptoms begin. Already, you can see in real time how your diet impacts your blood sugar, giving you a personalised road map for how to eat, although, as we discuss in food, eat real food. Recently, Samsung announced a watch and ring, featuring AI health monitoring, including FDA-approved sleep apnoea detection and diabetes tracking.

Others have discovered other type 2 diabetes (T2DM) glucose markers. Staphgen [240]measures the levels of an enzyme in the intestines called alkaline phosphatase (IAP) and says it can pick up T2DM earlier than the current gold standard Haemoglobin A1c test. They then make a supplement that restores that enzyme. So, expect a raft of new measures and medicines.

13.13.3 Aptamers and Other Measures

Future wearables will include smart jewellery, as well as wrist bands and electronic watches. But clothing is useful for large areas of measurement. Shirts, socks, shoes, patches (including the GCM) and gentle skin like sensors. Monitoring air quality will be included. We already have dogs that can smell cancer VOCs and as sensors continue to be more sensitive. Rapid antigen tests we got familiar with during COVID use the fact that if SARS-CoV-2 antigens are present in the sample, they will be captured by the antigen-specific antibodies and then seen as coloured lines on the strip test, indicating a positive test. The latest development for rapid testing is aptamers. Aptamers are oligonucleotides, such as ribonucleic acid (RNA) and single-stranded deoxyribonucleic acid (ssDNA) or peptide molecules that can bind to their targets with high affinity and specificity due to their specific three-dimensional structures.

Aptamers offer unique advantages over antibodies as molecular recognition agents, including for SARS-CoV-2. [241] Aptamers can achieve specific binding to target molecules with affinities similar to antibodies and used for the molecular recognition of their respective targets. They can be used both to treat disease or measure. Aptamers are still in the early stages of use, but because they can detect down to parts per billion (100,000 times less than ppm), they can be used for pathogen and cancer recognition, monitoring environmental contamination, and as stem cell markers. Companies are developing aptamer patient handheld devices that will be able to even detect prostate cancer and colorectal cancer. Imagine that in a few years, you would go to the doctor, prick your finger for a drop of blood and have a better diagnosis than a colonoscopy, or the yuck factor of doing a stool swab.

Dr Snyder and others believe that we should start out with a smartwatch, but the challenge is to get the medical fraternity on board. He suggested that there needs to be some financial incentives. Doctors may spend only 6 to 15 minutes in a consultation, whereas continuous monitoring can assess hundreds of measures with AI. It is just the start, and we will see a huge change over the next 5 to 10 years. For too long, health has been about trying to fix things up when you are sick. It is not focused on primary health care, which is preventing you from being sick in the first place. Or even if you are sick, catching it before you are seriously unwell.

13.13.4 Dashboards – Just Like a Car

While the research area is working out all the measures, a dashboard can summarise all of that complex data. All cars have a dashboard. Oil light, brake lights, doors open. A racing car may have up to 400 sensors, but they use a dashboard to summarise and can drill down to the information needed.

> *You don't drive a car around without a dashboard. I would argue it's just as crazy to go around without a health monitor.* Prof Michael Snyder

Health monitoring is similar, and currently, medical information is poorly organised. Information is often located in silos and is not easily accessible, or, as we discuss below, not yours! Some envisage a dashboard where, if a cardiovascular light goes off, the cardiologist can dig down to see which sensor triggered it and follow up on it. Many of the current measures and tests, such as pathology reports and imaging, are still poorly recorded and reused.

13.13.5 Skin Is Health

The latest research suggests that our skin is not just a mirror of our lifestyles – reflecting the effects of years of smoking, drinking, sun exposure, and stress, and hinting at our inner health. No, in this new upside-down world, the body's largest organ is an active participant in our physical well-being. This is a strange new reality where wrinkles, dry skin and sunspots cause ageing, instead of the other way around. [242] Weathered or unhealthy skin is emerging as a major risk factor for almost every single age-related disease, from Parkinson's to type 2 diabetes. The data from the Baltimore Longitudinal Study shows that participants who were judged to look at least 10 years older than they should were 99% more likely to have health problems. [243] Aged or diseased skin has been linked to the onset of cardiovascular disease, type 2 diabetes and cognitive impairment, as well as Alzheimer's and Parkinson's disease. Sun-damaged skin releases chemicals that contribute to systemic inflammation, increasing the risk of almost every age-related disease.

In a 3-year study, people had skin moisturising creams applied twice daily. The participants' cognitive functioning was measured at the beginning and end of the study. The control group had declined significantly, but those who had been hydrating their skin had not deteriorated. [244] Many natural ingredients can help. These include glycerol, petrolatum, hyaluronic acid and lipids that would usually be found in this layer of the skin – normal constituents of even the most basic moisturisers.

13.13.6 Ownership of Health Records

One unforeseen difficulty with dashboards is the ownership of your health records. In Australia, the health provider who created your medical record owns the record. [245] However, you have the right to access your health record. This right may be exercised in various ways, including viewing the record at the health service. When you request that your records be made available to another medical practitioner, that practitioner should act on that request but it turns out your previous doctor / lab / hospital / radiology is under no obligation to provide their original records. Further, the medical profession states these should be retained by the practitioner who created them. If you move from one medical practice to another, your data does not transfer with you unless you sign the appropriate forms. And generally, the case notes do not follow. If anyone has tried to get data you probably have struck the issue that health professionals can hold onto it tightly and won't make it available without a fight.

A similar process exists in the UK where patient records, both private and NHS, are not the patient's property and patients are not entitled to take possession of the originals. They can generally view the data. In the USA, it is more nuanced: in most states, the original physical medical record is the property of the physician's office that generated it, except in NH. Many think they own the record, but for the most part, they don't, and they can be charged a fee to produce a copy. Dashboards may take a while, given doctors' reticence, who owns the data and data privacy.

13.14 Drug Company Profits

Part of the discussion about health is that drug companies' values and profits do not align with societal objectives, but align with shareholder returns. One argument is that pharmaceutical companies need substantial revenue to fund research and development to create new drugs. Which sounds reasonable... until you look at the numbers.

In 2023, Pfizer only spent 16% of its revenue on research and development. It spent $3 billion more on marketing and administration than on research and development, yet still made $2.4 billion in profit. Johnson and Johnson spent $15.4 billion on research and development, $24 billion on marketing and administration, and still made a profit of $13.3 billion. This is replicated throughout the industry, with billions in profits being passed on to shareholders rather than being invested in the crucial development of the next wonder drug.

We are told that the development of drugs is funded by private funding. Nothing could be further from reality. The National Academy of Sciences [246] found that every single one of the 356 drugs approved by the FDA between 2010 and 2019 had received substantial public funding totalling $187 billion. The mean is $1.3 billion per target for basic research on drug targets and $51.8 million per drug for applied research on products. NIH investment was certainly no less than the industry's, with comparable accounting for basic and applied research, failed clinical trials, and the cost of capital or discount rates. Furthermore, companies benefited from funding from other government agencies, nongovernment organisations, academic institutions, and nonprofit organisations through contract funding, research and development tax credits, and vouchers. The authors concluded that this underestimated the public sector's contribution to pharmaceutical innovation.

The most important thing to remember is that the US pharmaceutical companies are focused on shareholder benefits. Not humanity. Not Tracey.

Certainly not Ada. They do not acknowledge that it is taxpayer funding that has helped them bring drugs to market. It is hard to stay optimistic. Keep that in mind as we look more into health advances.

13.15 Antimicrobial Resistance

Antimicrobial resistance (AMR) is a critical global public health challenge that threatens the foundations of modern medicine.

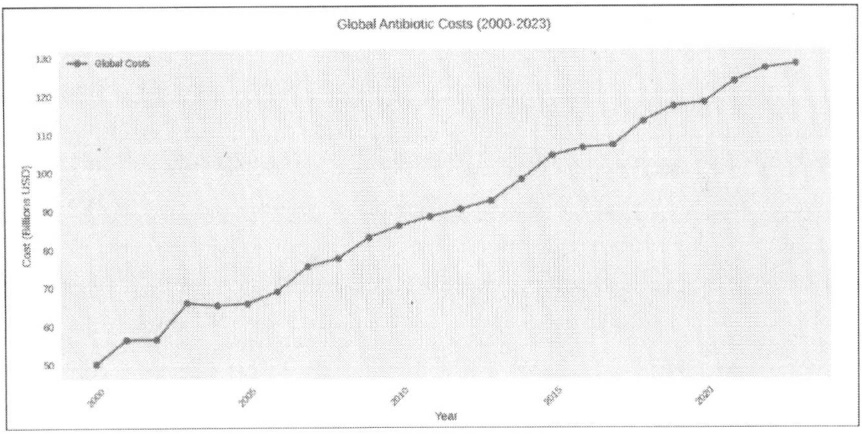

Figure 63 Increasing Global Antibiotic Costs 2000 to 2023

Figure 63 shows the increased AMR costs (billions) over the past 20 years. AMR occurs when bacteria evolve mechanisms that allow them to survive antibiotic exposure, making infections more challenging to treat. This resistance can develop naturally or be accelerated by the overuse of antibiotics. Resistant bacteria can spread from person to person, even without antibiotic use, and cause infections that are difficult to treat.

13.15.1 AMR Getting Worse

The problem is getting worse. Here are some facts about the issues.

- We have known about AMR for over a decade. [247]
- AMR incidence is increasing.
- Older age groups are more at risk.

- Costs have gone up 250%.
- Asia is the worst area for AMR.

Could you die from a simple dental infection? AMR means this could become an everyday reality. Bacteria and other microbes are evolving to resist the drugs traditionally used to kill them, such as antibiotics, transforming treatable infections into potentially deadly threats. AMR is often dubbed the "silent coming plague," jeopardising our ability to treat patients safely, crippling healthcare systems, and complicating care delivery for the most vulnerable in society. AMR is a growing global threat, responsible for one in eight global deaths a year (7.7 million), making bacterial infections the second largest cause of death globally. Out of these bacterial infection deaths, almost 5 million are associated with bacteria that have developed resistance to antibiotics, and 1.7 million are caused by pathogens resistant to all antibiotics.

That's simply inadequate. There has been a 65% increase in resistant infections from 2000 to 2023, with mortality rates reaching 4.95 million deaths annually attributed to AMR. Low- and middle-income countries show resistance rates 3-4 times higher than those in high-income nations. Economic impact analyses estimated annual global costs at US$100-150 billion, projected to reach US$300 billion by 2030. Age-stratified data showed a 2.5-fold higher risk in elderly populations and a 1.8-fold increase in pediatric cases. [248] is estimated to kill over 50 million annually by 2050.

Women may be more likely than men to develop drug-resistant infections – an underappreciated aspect of the growing threat of antibiotic resistance. The reasons for this gender imbalance are due to a number of reasons:

- Women have more urinary tract infections.
- Women get more antibiotics prescribed.
- Women face higher baseline susceptibility to urinary tract infections due to anatomical differences, hormonal changes, and sexual activity — and UTIs are also highly prevalent among pregnant women. This alone drives enormous antibiotic consumption. Women more exposed to antibiotics for recurrent urinary infections and reproductive issues accumulate a higher intestinal burden of resistance genes.

- Women make up 70% of healthcare workers, and they spend more time in health practices. E.g., drug-resistant Escherichia coli is one of many bacteria that women might be more likely to encounter than men in some regions, owing to these gendered divisions of labour.
- Genetics. A recent study led by the University of Turku, Finland, analysed the DNA maps of more than 14,000 gut metagenomes from a global dataset and found differences in antibiotic resistance between genders. In high-income countries, women had more antibiotic resistance genes than men. [249]
- Policy failure. More than 70% of countries fail to recognise gender inequalities in their national plans to tackle drug-resistant infections.

13.15.2 Agriculture to Blame for AMR

So how did this happen, and what is the connection between AMR and agriculture? Antibiotic use is much higher in animal production than in humans, with 70% to 73% of all antibiotics produced worldwide used in agriculture, primarily for livestock farming [250], including treating infections, preventing diseases, and promoting growth in animals. That 70% is over 80,000 tonnes in just the first 6 months of 2024, and is expected to rise to 200,000 tonnes by 2030. A recent Nature paper [250] [251] shows antibiotic use in farming is set to soar by 8% from 2020 to 2030, despite drug-resistance fears, and that antimicrobial drug use in agriculture is much higher than reported.

We have used antibiotics in animals for over 60 years, initially to improve growth rates. Although antibiotic use for growth was banned in 2006, over 40 countries still use them to improve growth.

In the USA, over 80% of all antibiotics are used in farm animals, not people. Animals get them by the container load; we get them by the small box. The data are not much better for Australia, which imports over 800 tonnes of antibiotics annually. Half goes into stock feed, 8% to veterinary use, and about ⅓ of a per cent to humans. Much of this use in animals is prophylactic (routine prevention) and enables farm animals, most often pigs and poultry, but sometimes also cattle, to be kept in conditions where disease spreads easily.

The overuse of antibiotics in farming contributes to higher levels of antibiotic resistance in some human infections. In the UK, antibiotic use in livestock has fallen 50% over the last five years, but in Asia, the Far East and Oceania, farm antibiotic sales per livestock unit are nearly four times higher than in Europe. For example, in pigs, antibiotics are given at weaning to reduce diarrhoea. The Danish FDA reported that 68% of pigs in conventional piggeries carry Methicillin-resistant Staphylococcus aureus (MRSA), whereas in organic piggeries, the percentage was 6%. In some countries, poultry farmers continue to use fluoroquinolones added to drinking water, even when no disease is present. Fluoroquinolones treat serious human Campylobacter infections, but are not used in the USA, Australia or Nordic countries.

13.15.3 How to Reduce Antibiotic Use?

The AMR problem demonstrates that the dollar is mightier than policy. We can see the problem. Manufacturers may see the problem, but they have quarterly revenue targets to meet. If not addressed, AMR will suck trillions from the global economy and put pressure on already struggling healthcare systems. Could the next pandemic be on the horizon? Scientists say the development of new antibiotics, vaccines, and diagnostics needs urgent attention. AMR means this could become an everyday reality. The WHO warns that by 2050, AMR could cause global damage on par with the 2008 financial crisis.

13.16 Mental Health

Mental health has not benefited from all the R&D in mainstream health. In Australia, over 3,000 take their own lives, nearly three times the number of road deaths. Despite additional funding, mental health takes its toll on the community. I lost a daughter to suicide over a decade ago, and the health systems appear paralysed about what to do. Treatment and care appear unchanged in the following decade, in contrast to the advances in physical health. A global rate of depression of 5%, and the fact that over 14% of American adults are depressed [252] is depressing news.

Australians like antidepressants [253] with about 1 in seven taking them and between 2013 and 2021, the antidepressant prescription rate steadily increased by 4.5% per year. Why? It appears that GPs prescribe antidepressants to patients experiencing distress even though the don't have a psychiatric diagnosis. Over-prescription of drugs is a worldwide trend.

The risk of all-cause, cardiovascular disease, and ischemic heart disease was 62%, 79%, and 121% higher, respectively, for those with moderate to severe depressive symptoms compared to those without depressive symptoms. Other studies have shown links between depression and dementia. Some would blame food. Some social media. But that is a story for another day.

13.17 Health Conclusion

We all want good health. Yet we all know someone who appears in good health, but then gets a diagnosis of breast cancer, colon cancer, dementia, a virus or a stroke. As we discovered in the food section, over 75% of food consumed is pre-packaged. Food can be medicine or poison to us. Many "health" services should be called "sick" services. The focus of health care over the past 100 years has been on providing care to the sick. Not to the well and not to prevent or head off illness. Polluting our home has been bad news and the consequences will be more cancers from the toxins we have created. Climate change as a result of burning fossil fuel is also going to impact billions of us..

We can use the analogy of a cliff or a waterfall. Is it better to stop people from falling from the top, or trying to fix them up after the fall. Current medical care focuses on treating people after they become patients (bottom of the cliff) rather than preventing illness (top of the cliff), leading to high costs in treating chronic and late-stage diseases. Additionally, a "one-size-fits-all" approach to healthcare does not account for individual differences in genetics, environment, or lifestyle, thereby reducing the number of people who benefit from interventions. Primary medicine is underfunded globally. Despite this underfunding, we have had notable successes such as vaccination programs for childhood diseases, including smallpox and polio, as well as for cervical and uterine cancers. Our existing and future diagnostic improvements and measurements promise to detect cancers at very early stages and improve outcomes.

We can be very optimistic about advances in health prevention and treatment. We are beginning to understand some causes that were previously unknown. That's due to advances in diagnostics and the power of AI enabling understanding of the interactions between the "omics" and we will explore AI more in the following chapter. Commercialisation of medicine moves slowly.

Most processes take a long time from lab to daily use. Incumbent corporations own our data and won't give it up without a fight. Diagnosis can be deployed rapidly, but medical lobbying keeps governments locked down. For example, a new treatment might be approved by the FDA/regulator, but the treatment is not deployed without a treatment code approved by the government or a private insurer. Solving the diagnosis is the first step, even if treatment takes longer. Treatments using multi-omics on a one-to-one basis are generally very expensive ($Ms), but there is hope (not hype) that reductions in the costs of new technologies will bring treatments to all.

Tracey can look forward to better care in old age. Ada – her primary concern is having enough food and energy for today. Tomorrow is another problem.

Conclusions from Health:

- Even without AI, our understanding of health is improving – progress on cancer and heart disease, though dementia remains stubborn.
- Exercise continues to be the most useful preventative action for health.
- There are early suggestions we can live long (120+ years) and die young (without disease).
- Measure more to understand small changes – early detection changes outcomes.
- Man-made chemicals and plastics – particularly the 2 credit cards' worth we ingest – will have long-term health consequences.
- Mental health advances are slow and poorly funded – expect a worsening over the next decade.

---

14 AI for Better Health

AI for Better Health:

- How will AI improve health outcomes?
- Understand how AI may provide cheaper drugs and therapies.
- Will the medical fraternity allow AI to bring improved diagnosis to 6.5 billion people?
- How does mental health fit in this new AI world?
- Advances in brain-machine interfaces.
- How AI is speeding up clinical trials and mapping diseases like cancer at a cellular level (the Human Tumour Atlas).

Artificial intelligence is providing tools in multiple areas of health, and we will explore a few to understand the far-reaching implications to modern health - from research to clinical treatments. AI will probably be the most important technology in health. It will enable progress in every sector of health – but will the gatekeepers slow progress? How will advances affect Tracey in Western countries or Ada who lives with the 6.5 billion without good health care?

AI and machine learning are transforming the healthcare industry, which accounts for 11% of global GDP or $9 trillion annually. Healthcare is over 20% of GDP in the USA, despite being 47th on the country's health score and paying three times more for health than other developed countries. AI is transforming drug and vaccine development and improving medical diagnosis and treatment. AI technology is being used at all stages of the value chain, boosting efficiency across the healthcare system. [254]. Google has announced

MedGemma, a 4 billion parameter multimodal model and a 27 billion text-only model for medical text and image comprehension which will enable them and others to advance diagnosis and treatment.

The healthcare (or sick care industry) is slow to change. My nephew in his PhD showed machine learning using Eigenimage analysis on x-rays was better than radiologists picking up lung cancers in 2004. [255] Two decades later, Kim [256] shows accessible and adaptable artificial-intelligence model trained on a diverse set of X-ray images is better than existing models at spotting rare chest diseases. About half of breast cancer screening services globally now utilise a New Zealand-developed AI to provide faster and more reliable patient feedback, but this has taken decades. There are still many breast screening health services globally that still do not use the technology, despite the demonstrated benefits to both clinicians and women.

LLMs that deal with hundreds of thousands of words can deal with the simplicity of RNA, DNA and proteins without much effort at all. The data set is much larger given the human body is made up of about 30 trillion cells and an equivalent number of bacteria. AI is important in healthcare for many reasons but the main focus should be "*healthcare systems to be smarter, faster, and more efficient in providing care to millions of people worldwide*". It needs to focus on keeping people well, rather than letting them get sick. AI can be trained on all the medical information from decades of prior information as well as sifting through new information. We will see an increasing use in health-enabling multi-omics, diagnosis, drug development, medicine use and medical education. More importantly from both an economic and a patient perspective, AI will prevent illnesses and stop you getting sick.

The field is moving so fast that this book will be out of date within months, so if you are going to follow AI and health, you will need to follow other media than printed material. For example, in February 2025, the Arc Institute and Nvidia announced [36] an open-source AI for tracking interactions beyond single-cell genomes of bacteria and archaea. It will include information from humans, plants, and other more complex single-celled and multicell species in the eukaryotic domain of life.

Early validation of the Evo 2's capabilities showed that this model can identify how genetic mutations affect protein, RNA, and organismal fitness. In tests with variants of BRCA1, a gene associated with increased risk of breast and ovarian cancer, Evo 2 achieved greater than 90% accuracy in predicting which mutations are benign versus disease-causing. Open source means researchers everywhere can tap into the data, so expect more to come of this over the next few years.

14.1 AI: Better Diagnosis

Google Research and Google's AI research lab, DeepMind, have detailed the impressive reach of Med-Gemini, [257] a family of advanced AI models specialised in medicine. It's a huge advancement in clinical diagnostics with massive real-world potential. [254] To deliver effective care, they must be familiar with each patient's health record and up-to-date with the newest procedures and treatments. And then there's the all-important doctor-patient relationship, which should be built on empathy, trust and good communication. For an AI to come close to emulating a real-world doctor, it must be able to do all these things. Med-Gemini was trained on MedQA, multiple-choice questions representative of US Medical License Exam (USMLE) questions designed to test medical knowledge and reasoning across diverse scenarios.

Med-Gemini is better than SOTA or State of the Art experts. It outperformed humans for:

- Medical summarisation
- Referral letter generation
- Medical simplification tasks
- Better at communicating outcomes.
- Keeping up with the latest research and best practice treatments
- Avoids information overload

Medically as AI becomes be fine-tuned it could to provide high-quality long-form answers to nuanced and open-ended medical questions which are asked by millions of internet users. Med-PaLM 2 surpasses physicians on axes such as factuality, reasoning, harm, and bias.[258] The authors provide an example of a conversation with med Gemini in this following image.

I have provided a LLM with my 15 years of blood pressure, lipid profiles, surgeries and the outcome is better questions than my local GP.

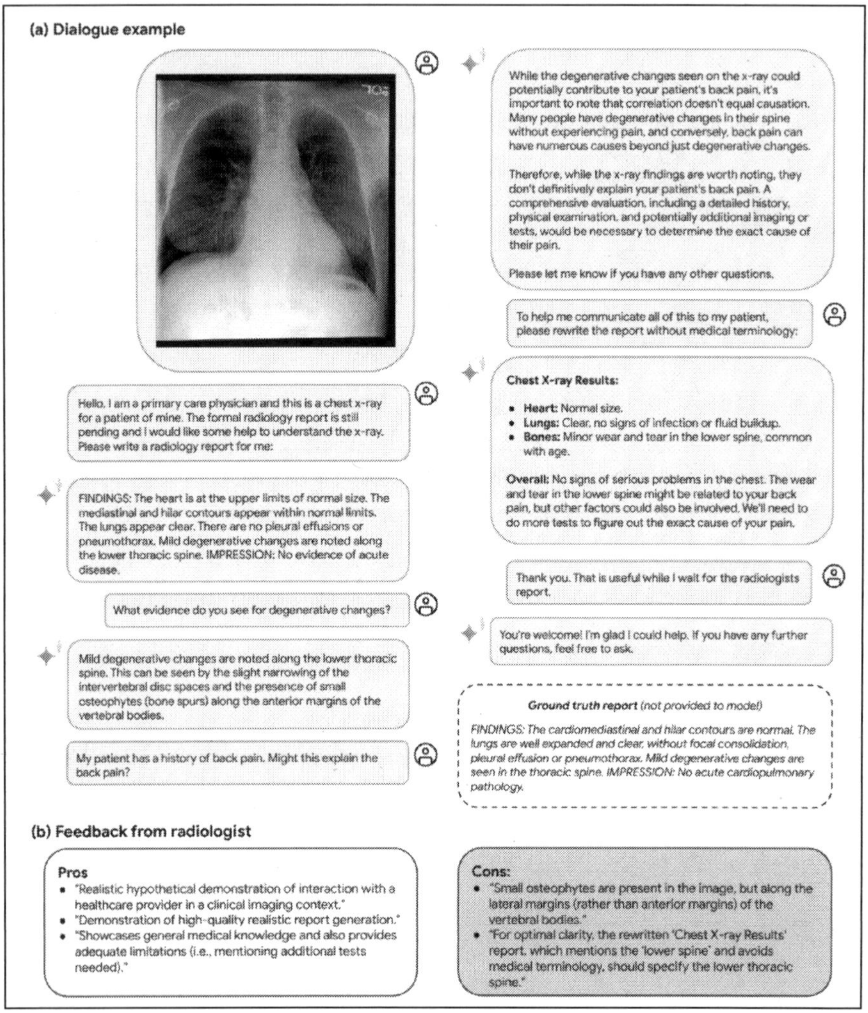

Figure 64 An example of a hypothetical doctor and x-ray for a patient.

In Figure 64 Google provides an example of a conversation between Med Geminia and a worker.

- Med-Gemini-M 1.5 analyses a chest X-ray (CXR)
- Has realistic dialog as if you were talking to a physician.
- Demonstrates promise. Identifies mild degenerative changes along the spine and can answer questions about the reasoning which led to

this finding, demonstrate general medical knowledge about degenerative disk disease and distinguish between correlation and causation in relation to a patient history of back-pain.

- Creates a layperson answer, helping patient understanding and communication

The radiologist feedback is illuminative who states "*minor changes only*". We know there are numerous instances where thousands of patients have been misdiagnosed. An AI system remembers all the x-rays viewed. It does not get cocky or is rushed off its feet, nor does it come into work hungover from a night out with colleagues. The evidence is clear, yet it not fully adopted.

The price of a diagnosis is falling and the models are rapidly improving.

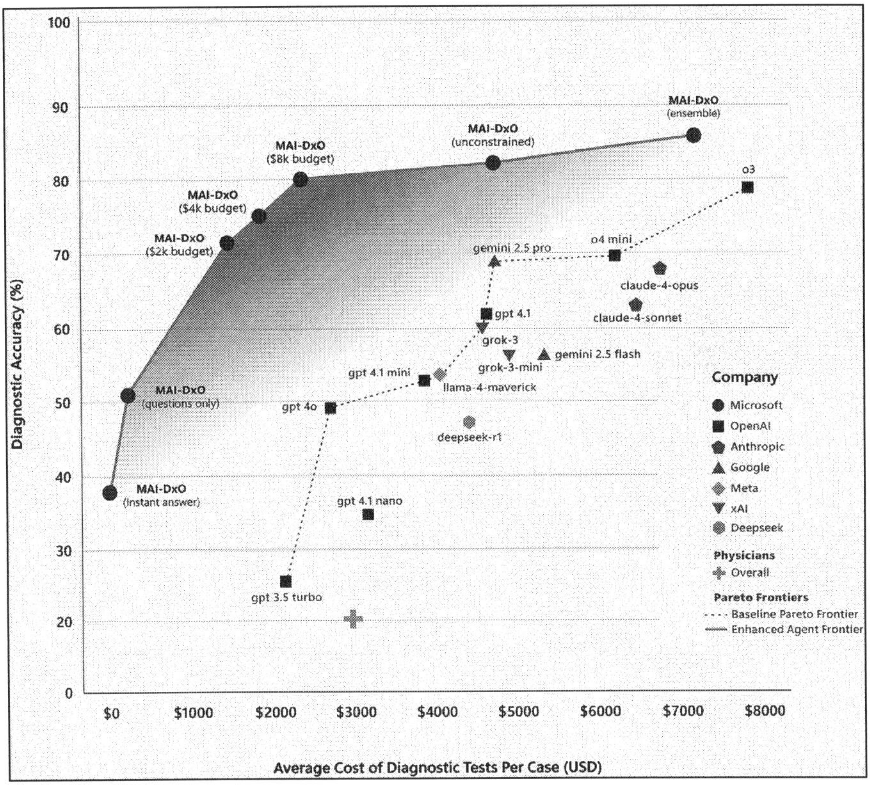

Figure 65 Microsoft OpenAI Diagnostic Accuracy vs Costs

Figure 65 shows the average cost of diagnostic tests per case using a number of AI models including Microsoft, Google, X, Meta, Anthropic and Deepseek. The physicians average cost was $3,000 but with a diagnostic accuracy of only 20%. In contrast Microsoft / AI framework is four times better than doctors, [259] but costs up to $7,000. They benchmarked these against real-world case records published each week in the New England Journal of Medicine and stated the Microsoft AI Diagnostic Orchestrator (MAI-DxO) correctly diagnoses up to 85% of NEJM case proceedings, a rate more than four times higher than a group of experienced physicians.

MAI-DxO also gets to the correct diagnosis more cost-effectively than physicians. Some claim this result is misleading as they used zero healthy patients. Doctors were banned from googling symptom, consulting colleagues, using up to date medical databases or calling specialists. The AI was never tested on "this headache is just stress", "Let's wait and see" or "More tests will cause more harm than good". A comment from a doctor said the benchmark rewards finding zebras, not recognising horses. Given the misdiagnosis that I personally have received, anything to improve outcomes is valuable.

14.2 AI: Cheaper, Better Drugs

Advances in fundamental biology, artificial intelligence, automation, and trial design should lower preclinical drug development costs significantly. They enable methods that eliminate less-promising candidates early in the drug development process, prevent downstream misallocation of R&D capital, and create a larger chemical search space early in the discovery phase. During the next decade, companies leveraging these techniques fully could lower costs per approval by almost 50%, in part by more than doubling the odds of success for those drug candidates that do enter clinical trials. Dr Snyder is the author of the book: "Genomics and Personalised Medicine: What Everyone Needs to Know".[260]

14.3 AI: Better Clinical Trials

As OPYL [261] points out, clinical trials stand as the cornerstone of medical advancement, a necessary step for any new drug or medical device before it is brought to the market. Yet, traditional methods employed in designing these trials have remained stagnant for over three decades. Using ChatGP and 3 years of machine learning and big data analysis they have just released an AI product called Trialkey.ai [262] with over 400,000 clinical trial data from the repository at clinicaltrials.gov. Much of the data was free form but they have developed over 800 data variables. They matched where possible following approval through FDA and commercialisation. That data was often not available - phase 1, 2 or 3 may have been abandoned, completed, amended or not progressed or commercialised. Despite these limitations data ai can predict to 90% whether the objectives of trials were met or not.

The ramifications are enormous. The system virtually provides over 30 variables to game trial design and ensure successful clinical trials. Users can run virtual trials and optimise these to get better outcomes. Trialkyl expects as it is further developed that the 30,000 clinical trials each year will result in better outcomes, that more money is put into fewer trials (as low as 6,000) with higher probability of outcomes and reducing trial times. In particular, government and university trials have poor outcomes. Listen to a great podcast on Trialkey founders on Ey-on.au podcast episode 184. They are not the only ones. OpenAI and Moderna have teamed up and Moderna is leveraging OpenAI technology to augment clinical trial development by building an AI to help analyse large data sets. Colour Health built an AI CoPilot using GPT-40 to assist healthcare providers to make evidence-based decisions about cancer screening and treatment. Vitara Health in conjunction with breast screen services globally uses AI to provide women with a rapid response following breast screening. A German and USA study using AI looked at nearly 1.2 million mammograms and found that having a radiologist and AI system working together was 2.6% better at detecting breast cancer than a radiologist alone, and in another study shows AI assisted may reduce unnecessary testing. [263] In that study, out of 10,000 who underwent exams, 262 could have avoided diagnostic exams, 10 avoided biopsies, without any cancer cases missed. Workload would be 24% less.

The cost of new drugs is about $2.1 billion, and it takes about 13 years to develop a new drug from the start. That cost includes the costs of drugs that never make it to market, whether pulled at early clinical trials or, in some cases, just before final release. Companies have to make over $1 million a day to warrant that investment, and when you hear of drugs costing hundreds of thousands a year, that is due to the whole drug development cycle. Drugs developed with AI in 2019 are now coming to market with a total investment cost of about $700 million and entering the market in less than 6 years. That is a 12-times improvement in drug development. That means that the return required is substantially less, and we will see new drugs for diseases that were too small a market to warrant the investment.

14.4 AI: Human Tumour Atlas

The human tumour atlas, HTAN, is a National Cancer Institute (NCI)-funded Cancer Moon-shot initiative to construct 3-dimensional atlases of the dynamic cellular, morphological, and molecular features of human cancers as they evolve from precancerous lesions to advanced disease.

- Humans are loaded with cancer cells.
- Most don't cause cancer.
- Our immune system keeps them in check.

That is part of the reason why more older people get cancer. The immune system becomes less effective with age. The environment also keeps the cancer from growing. Ugur Şahin and Özlem Türeci, who co-founded BioNTech, the German firm that partnered with Pfizer to manufacture a revolutionary mRNA Covid vaccine, said they had made breakthroughs that fuelled their optimism for cancer vaccines in the coming years and possibly by 2030. [264] We should expect rapid improvement in cancer detection and prevention due to rapid improvements in measuring and monitoring. It is looking for a needle in a haystack, and without big data and AI impossible for humans to find. AI is a tool to simplify this search, and as more applications for AI are proven, the more likely we are to find those needles, but it is probably 2035, not 2030...

14.5 AI: IBD

Inflammatory or irritable bowel disease (IBD) and related conditions including Crohn's disease cause suffering and ill health to millions. [265] An estimated 3.2 million to 13 million people in the UK have IBS. Because many individuals experience symptoms but never seek a formal diagnosis, exact figures vary. The NHS and Guts UK report that IBS affects roughly 1 in 20 people (about 3.2 million individuals), but many health organizations, including Bupa UK, estimate that up to 1 in 5 adults in the UK (over 10 million people) may suffer from the condition. A major trigger of millions has been identified using the latest genomic techniques and using AI.[266] Moreover, AI enabled the researchers to identify there were existing drugs (MEK inhibitors) that could stamp it out, in what scientists call a "massive step" in successfully treating these debilitating chronic conditions. Until now, our incomplete knowledge of what drives IBD has made it incredibly hard to find a 'one-size-fits-all' medical intervention. As the authors point out

> *"Together, this illustrates the power of functional genomics, applied directly in primary human cells, to identify immune-mediated disease mechanisms and potential therapeutic opportunities."*

Irritable bowel disease, IBD, AI can lead to our understanding of disease. Today genomics, part of a larger movement toward personalised medicine, is poised to revolutionise health care.

14.6 AI: Muscular Dystrophy

We are at the tip of the iceberg in understanding the causes of disease. As reported in Nature in mid-2024, Duchenne muscular dystrophy should be a prime target for gene therapy. While scientists have found some success in the past, a major hurdle is that the gene that encodes dystrophin is one of the largest known, so it's far too large to package into the viral vectors typically used to insert healthy copies of genes into cells. [267]. Scientists have found a way, and human trials are likely within 2 years.[268] The Australian Government is investing $27 million in 10 AI-driven projects to improve the diagnosis of genetic diseases, including multiple sclerosis, skin cancer and cardiac disease, through the Medical Research Future Fund. Dozens of other countries are doing likewise, and a paper back in 2019 [269] showed the limited extent. Five years on, the Research Director of CSIRO's Australian e-Health Research Centre says we're at the cusp of an extraordinary era in medicine. For the first time, machines can provide efficient administrative support for clinicians, patient education, new diagnostics, new treatments, disease prediction, and informed clinical decision-making. [270]

14.7 AI: Vaccines

As pointed out previously, vaccines have played a key role in reducing disease. We have over 35 vaccines in use and a further 15 pipeline vaccines or monoclonal antibodies (mAbs) in development [271], including malaria. Child vaccination is one of the most cost-effective health interventions. As we have previously discussed, vaccination in the major driver for reduction in infant deaths since 1974 and has led to big health gains over the past two decades. [272] Vaccines have saved over 154 million lives over the past 50 years, with 60% of those from measles, and 95% of children are vaccinated under 5 years.

Progress and regression are common. In Australia in 1995, only 52% of one-year-olds were fully immunised. By 2020, Australia had reached 95% coverage for one-year-olds and five-year-olds. Regrettably, by 2025, child vaccination rates are falling back under 90%, with warnings of the re-emergence of childhood diseases such as measles, whooping cough and other childhood diseases.

In the USA, more than 1,000 cases of measles have been confirmed since January 2025, including a cluster in West Texas that has caused one of the worst outbreaks. These outbreaks are occurring even though measles was technically eliminated in the U.S. back in 2000.[273] Most are in schools operated by Christian evangelical churches.

Worldwide, more than 107,000 people died from measles in 2024, despite the availability of an effective, safe, and inexpensive measles vaccine licensed since 1963. The majority of these deaths occurred in sub-Saharan Africa, where vaccination rates were only about 70%, with COVID and conflicts disrupting vaccination programs. That is less than the 140,000 in 2018, but it shows the challenge is social, not technical. Even the European region is not immune, with over 127,000 cases of measles in 2024. [274]

mRNA technology introduced during the COVID vaccine development enables new vaccines to be more effective, faster to market and more specific. With new technologies, we should expect to see more effective vaccines over the next few years, not just for existing diseases but also for cancers.

14.8 AI: Brain Computer Interfaces

Several researchers and companies are looking to implant invasive brain-computer interface systems (BCI) aimed at regaining or restoring walking ability in human individuals with severe injuries including spinal cord injury

(SCI) or genetic conditions causing loss of ambulation. About 15 million people globally live with SCI. [275] Many restrictions in performing activities and participating in meaningful life do not result from SCI itself, but from other contributions due to insufficient or inadequate medical care, rehabilitation and access to assistive technologies, a high economic burden, and physical, social and policy barriers. For example, globally, only 5–35% have wheelchair access. Organisations such as Australian Wheelchair for Kids send wheelchairs to children globally, but they produce about 300 per month.

Company/Research Group	Key Focus/Technology	Mobility Restoration Application
BrainGate (Brown University, Massachusetts General Hospital, Stanford University, University of Pittsburgh, among others)	Intracortical microelectrode arrays (e.g., Utah array) to record neural activity.	Enables paralysed individuals to control robotic arms, computer cursors, and other assistive devices with their thoughts.
Blackrock Neurotech	Implantable brain-computer interfaces (BCIs), specifically the NeuroPort Array.	Restores tactile function, movement of limbs and prosthetics, and control of digital devices for individuals with paralysis. Their MoveAgain BCI system has FDA Breakthrough Device Designation for immobility.
Synchron	Endovascular BCI (Stentrode device) is implanted in blood vessels, avoiding open brain surgery.	Deciphers neural code to restore motor intent and control digital devices for people with limited mobility.
Neuralink	Flexible, high-density electrode arrays (N1 implant) and a surgical robot (R1) for precise implantation.	Initial human trials focused on restoring mobility and communication for paralysed individuals (PRIME study), with long-term goals of enabling thought-controlled devices.
Onward Medical	ARC-IM (implantable BCI) and ARC-EX (external spinal stimulation) systems.	ARC-IM aims to decode brain signals for movement, potentially enabling standing, walking, swimming, and cycling for severe spinal cord injury. ARC-EX focuses on restoring hand strength and sensation.
Battelle	NeuroLife® Technology involves wearable sleeves that measure nerve and muscle activity. Bioelectronic Medicine program.	Recreates dexterous hand movements through targeted stimulation interventions, bypassing damaged pathways.
Northwell Health's Feinstein Institutes for Medical Research	"Double neural bypass" system using brain implants and AI.	Restores both movement and sensation in a paralysed hand by rerouting signals from the brain and providing sensory feedback.

Figure 66 Invasive Implants for regaining mobility

Figure 66 lists the various groups globally, their key focus and their mobility restoration application. Helping people move again through brain compute interfaces is really difficult. The understanding of the brain and how it communicates with the body is really complicated. The history of BCI is long and slow.

- Early years: Monkey/rodent studies (2016–2018) consolidate core principles including intent decoding, CPG targeting, and closed-loop circuit restoration.
- 2019. Translational efforts begin with first-in-human ECoG BMI.
- 2023. Clinical feasibility explodes: robust, chronic, natural gait restoration in a single human, bi-directional and sensory-augmented BCI systems.
- 2025. Emphasis is shifting toward multi-site, bidirectional, chronic human trials, driven by hardware advances and foundational animal work.

Neuralink [276] has three patients in clinical trials and uses a medical robotic device to implant its device and over 64 superfine threads containing 32 electrodes for 1,024 electrodes. This is significantly more than the 100 electrodes used in a Utah Array. While BCIs have existed for decades, Neuralink, owned by Elon Musk, has brought renewed public interest to a technology that could significantly improve the lives of those living with quadriplegia and others with disabilities or neurodegenerative diseases.[277] Neuralink and other companies appear to be expanding, but where AI and robotics will contribute is unclear. We could envisage that within a decade, this BCI approach could bring independent movement to some of those 15 million spinal cord injuries globally. Assisted with humanoid robots, we could imagine that the burden and cost of care for SCI will decrease and quality of life increase.

14.9 AI: Health Conclusion

We all want good health. What AI provides is the ability to comb or knowledge of the genome, biome, and other omics to find needles in haystacks. It will allow better diagnose, understand disease better and develop better drugs. Furthermore, inequality to healthcare means an estimated 6.5 billion people lack access to doctors and medical diagnosis. The AI from Google and Microsoft shows the power of improved diagnosis, but will the medical profession encourage or block such innovation? Of all AI sectors, will the gatekeepers, the old clinicians, allow this radical tool to deliver what early work has promised? For some disciplines such as drug development, AI has already made massive improvements. In other medical fields, it has scarcely made any mark at all.

AI for Health Conclusions:

- AI has shown that drugs can be developed for a 12 times better return, cutting time from 13 to 7 years and slashing costs from $ 2.1 billion to $600 million.
- Advances in AI will also provide health care systems with streamlined AI processes. While hospitals are slow to change,new ways of doing things are already underway.
- AI promises much in the interpretation of normal clinician visits, asking more questions and helping physicians and specialists.
- AI has slashed the costs of new drug development.
- AI will enhance vaccines.
- AI will revolutionise the medical care system, very slowly at first, but as AI matures, it will move into other systems to improve efficiency, create closer doctor-to-patient relationships and enable health to keep up with the increase in population.

---

15 What Do People Do?

What do People Do Without Jobs?

- Jobs will go.
- What do people do without jobs?
- How will economics and political will shake out?
- Why this disruption is different: previous waves displaced either cognitive or physical labour – this one displaces both, at once.
- Care work, the arts, and education – the jobs AI is least likely to replace, and whether they'll finally be valued properly.
- Universal Basic Income and sovereign wealth funds – could AI itself fund the transition for displaced workers?

With tasks taken over by machines for both service jobs and blue-collar manual tasks, what do people do? Do we become a nation of carers, artists and writers? This is one of the genuinely interesting civilisational questions of our time. The historical parallel isn't reassuring.

Every previous automation wave displaced categories of work but created new ones. The loom killed cottage weavers but created factory workers, mechanics and merchants. The tractor emptied farms but filled cities. This time the argument is different: AI and robotics are displacing cognitive and physical labour simultaneously, which is the first time that's happened. There's no obvious next sector to absorb the displaced.

15.1 Artists and carers

Care work is genuinely robot-resistant, at least at the relational level. People want human presence when they're dying, raising children, navigating grief. That's real and durable. But care work has historically been underpaid precisely because it was feminised and abundant. If it becomes economically central, it either gets properly valued for the first time or gets commodified and underpaid at scale — and which way that goes is entirely a political choice.

Art is trickier. AI already produces competent music, images, and prose. What survives is human provenance, the value of knowing a person made something. That's real, but it's a niche premium market, not a mass employment base.

The real issue isn't what people will do because humans are extraordinarily creative at finding purpose. The issue is how economic value gets distributed when labour is no longer the primary input to production. If robots generate the wealth and capital owns the robots, you get a hyper-concentrated aristocracy with a restless, purposeless majority. That's historically very unstable.

The serious answers cluster around three possibilities:

- **Universal Basic Income** — decouple survival from employment, let people self-organise purpose
- **Radical redistribution of ownership** — robot dividends, sovereign wealth funds, worker ownership of automation
- **Deliberate inefficiency** — societies *choose* to keep humans in roles even when robots could do it cheaper, as a social contract

A post-scarcity society could genuinely become one of artists, carers, craftspeople, teachers, community builders, athletes, and philosophers — the things humans do when survival is no longer a concern. The ancient Athenian citizen class lived roughly that way (on the backs of slaves, uncomfortably). The question is whether democratic societies can engineer that transition without the brutal inequality and social dislocation that inevitably precede it.

Given my 50 years watching the energy transition move at glacial speed despite the physics being obvious, I predict the political economy will lag the technology by decades, and the suffering in that gap is the real problem.

15.2 Education

What is the role of education if we do wind up in 2045 with humanoid robots (machines) capable of doing all jobs? Not only those that people currently do, but also jobs where the cost of doing them is too expensive today? Looking into the future is difficult. We don't know the exact timeframe because the emerging technology isn't yet mature enough to know which jobs, in which sectors, will be affected by when. We will know more in a decade as some tasks and jobs get replaced. When dirty, dangerous, and drudgery jobs are gone, what is the role of education in preparing you for a job that no longer exists?

If education is to prepare people for boring jobs, what is the role of education if those jobs are no longer there? Why would you get educated for a job? Would you get educated just because you can? This is a challenging concept to most. We have heard Prof Geoffrey Hinton say that the young person ought to train to be a plumber. But if 90% of that plumbing job can be done by a humanoid robot, why be trained?

15.3Numbers Drive Politics

The numbers cited previously are the crux: robot labour at $0.50/hr versus human labour at $15–35/hr, capital costs of $20,000 per robot, and robots making robots. The wealth doesn't disappear – it concentrates catastrophically fast into whoever owns the capital stack. The existing 3,500 billionaires figure is the starting condition; without intervention, it gets dramatically worse.

15.4 Universal Basic Income

Replacement of service jobs (AI) and physical jobs by humanoid robots (AI plus mechanical) has two impacts. What will people do without purpose, and what to do without income? Many thought leaders suggest that the value of a humanoid robot should not just benefit the robot's manufacturer or to the business that uses it, but to society as a whole. It is vital to protect people, not jobs, firms or industries.

- Jobs and Purpose. For some workers, their identity is closely tied to their job. When asked what is important, their responses often revolve around their careers. We have expressions such as "I have never worked a day in my life – my passion is my job." However, many individuals would prefer not to work but to pursue their interests—art, music, hobbies and crafts, restoring cars, gardening, and volunteering. Note that of the 4.2 million retirees in Australia (15% of the population), more than 56% of those over 50 are engaged in part-time work, either paid or voluntary. The main reason? Stay busy and keep their minds active.
- Jobs and money. Like the chant of the 7 dwarfs in Snow White, "I owe, I owe, so it's off to work I go." A money purpose often revolves around paying the mortgage or rent and may serve as a reason for working. A few opt out, but unemployment benefits are below a living wage in most societies.

A Universal Basic Income, or a Universal Prosperity Income is one solution to protect people but there may be other strategies. A tax on the humanoid robots is a simple solution. The humanoid robot can work 168 hours per week. That is four times the 40 hours that humans work, and it works without holidays, sick days, bathrooms, public holidays or any of the on-costs. A simple split for the additional value of the humanoid robot could be:

- 25% of the value goes to the robot manufacturer.
- 25% to the manufacturer.
- 25% to society.
- 25% to the humans whose jobs are taken over by the humanoid robot.

This split of value provides for society's health, education and social services of the society at large. In other words, protect people, not jobs, firms, or industries. How much should the laid-off worker get, and how? We need to start thinking about an equitable split now. AI and humanoid robots are here today, and it won't be long before they are everywhere. As AI and humanoid robots improve task performance, demand will outstrip supply. Some humanoid robot companies are focused on the consumer market, some on the business. Robots are already in the aged care sector. We need to get politicians with a clear direction to encourage humanoid robots AND make sure that all of society benefits.

15.5 Sovereign Wealth Fund

The idea is sound in principle and has real precedents:

- Norway's Government Pension Fund is the cleanest example – oil revenues redirected into a $1.7 trillion fund that now pays dividends to citizens.
- Alaska's Permanent Fund pays every resident an annual dividend from oil royalties.

Both demonstrate the mechanism works. The question is whether democracies can apply it to AI and robotics at the required scale.

15.6 Obstacles to Sovereign Funds

There are serious obstacles to sovereign wealth funds.

1. **Taxation jurisdictions**. The companies that own the robots – Tesla, Microsoft, Amazon, the Chinese equivalents – are globally mobile. They will domicile profits wherever taxation is lowest. Any single nation that imposes a "robot tax" faces capital flight. This requires coordinated international action of a kind that has largely failed on corporate tax for two decades.
2. **Speed mismatch**. As our S-curve slide implies, the disruption arrives in a 5–10-year window. Political systems operate on 3–4-year electoral cycles with 10–15-year policy implementation timelines. The displacement will be well underway before the policy response is designed, let alone enacted.
3. **The capture problem**. The people who own the capital are the same people who fund political campaigns, own media, and employ lobbyists. The historical record is not encouraging – the tax systems of every advanced economy have become progressively less redistributive over the 40 years that neoliberal orthodoxy has dominated, precisely during the period when the data showed inequality accelerating.

15.7 A Functional Model

If we were designing it from scratch, the architecture would probably need:

- A robot/AI productivity levy – a percentage of the value-add from automated labour, collected at point of production, not profit (harder to avoid).
- A national AI sovereign fund – ring-fenced from general revenue, managed independently like a central bank.
- Universal dividend payments – direct to citizens, not means-tested, avoiding the bureaucratic welfare trap.
- International coordination – probably through something like the OECD minimum tax framework, extended to automation

15.8 Hurdles

Companies such as Microsoft, Meta and Tesla are all "sucking up to the Australian Government, which represents exactly the leverage point where a smart government could negotiate equity stakes or productivity levies as the price of market access and infrastructure support. Singapore has done versions of this with sovereign wealth (Temasek). Australia has the minerals, the land, the sun, and the stable institutions but currently seems to be offering the red carpet for free.

The mechanism exists. The precedents exist. The political will, at the scale and speed required, almost certainly does not exist yet. What typically creates that will is a crisis visible enough and painful enough that the public demands it. The Philippines call centre story as we talked about in earlier section on the impact of AI destroying an industry is the leading edge of that crisis arriving. When it hits at scale in Western electorates, the politics will shift but probably after considerable damage.

15.9 Conclusion

We need a framing of "no one left behind". It is the right moral foundation. The harder question is whether it becomes policy before or after a generation of disruption.

Concluding: What Will People Do?

- We will not become a nation of carers and artists.
- The technical and economic solution is outstripping the political direction.
- A Sovereign Wealth Fund to capture on a national basis works, but not one yet has resisted the technology billionaires.
- Care work and the arts may become economically central, but whether they're valued or commodified is a political choice, not a given.
- Unlike previous disruptions, there is no obvious next sector ready to absorb the displaced workforce.
- This is one of the great civilisational questions of our time, and the answer is still ours to decide.

16 Conclusion - Overall

Overall Conclusion

- AI provides massive change across multiple disciplines – bringing both optimism and hard political questions.
- Healthcare advances – faster diagnosis, cheaper drugs, and a possible end to the worst of cancer within two decades.
- The end of toil and drudgery – how AI and robotics roll out in phases across the next two decades.
- Food disruption and the regreening of the planet – what happens to the land we free up.
- The 1,800 C-suite executives and 20,000 politicians who hold disproportionate power over our destiny.
- Practical steps for the transition across energy, health, food, labour, and politics – and what your part can be.

As we look at the world through Tracey's or Ada's eyes, we could easily be very despondent. For me, despite dozens of books, hundreds of references, and thousands of social media posts, my initial optimism remains regarding the end of toil, the end of food shortages, reduction in land use for agriculture, and advances in health. Most importantly, new technologies help us regain Earth's sustainable carrying capacity. We don't have any other home. Musk may or may not create a bolt hole on Mars, but I have zero confidence in that, too little and too late for me. Tracey might justify her optimism for the planet's future, whereas Ada, trapped in poverty, only wants the day to end.

We can be optimistic about the future from a technology or economics perspective. But can we be optimistic about those in politics or large corporations who actively resist change? We cannot trust those 180 fossil fuel companies that have caused pollution on our Earth. Each of those will have about 10 people in the "C-Suite" (CEO, CFO, CTO, etc). Those 1800 people have a powerful hold over about 20,000 democratically elected politicians (plus a few hundred non-democratic politicians). Our destiny is determined by this C-Suite and parliamentarians. I find it outrageous that those 1,800 people are destroying the home we live in. They are enabled by minders, engineers, and economists. Climate change is a real existential threat that we have the technologies to address today. AI superintelligence is a theoretical threat.

Those 1800 won't change willingly or rapidly, as they have clearly demonstrated. Like the tobacco industry, the fossil fuel industry has known the consequences of burning fossil fuels for at least 30 years. Like an elderly friend, I can speak to him and encourage him to come along to the gym 3 times a week for just half an hour. The evidence for a reduction in dementia risk is compelling. His risk of dementia is cut by a third. But will he do it? No. What stops humans from seeing the big picture? Some people don't want to think about the future because their life is a struggle, and problems in the wider community are overshadowed by their day-to-day survival. There is a disconnect between enjoyment and prosperity today, and some uncertain future consequences. We understand younger people think of themselves as invincible. They play sports with a high risk of head injury or drink alcohol like there might be a shortage tomorrow. But in the energy sector, we have mostly men, primarily white, many of them with children and grandchildren, but they don't seem to care about the future. Are health, food and agriculture industries also resistant to change? The answer is yes. Established industry titans are the same whether in energy, food or healthcare.

16.1Advances in Healthcare

The private and public health sectors in most countries focus more on treating the ill. We also see examples of less scrupulous or morally bankrupt making money out of victims. A history of snake oil salesmen and public sector health failures has led to a culture of cautiousness and a do-no-harm approach amongst health practitioners. The barriers to entry for medicines and medical devices, the high cost of conventional medicine, a lack of affordable access, and suspicion that conventional health is profit-driven rather than people-driven are factors that have turned people to the complementary health market. The complementary health market has ballooned to a $180 billion "wellness" market, 3 times the size of the pharmaceutical industry. This is not to dismiss some of the benefits of "natural" medicine. As we begin to understand the advances of the multi-omics revolution, it is easy to see that within 2 decades, we can see the end of, or at least a significant reduction of, cancer cases. We have evidence that cancers, which once were the disease of aging, are now rising in 30-to 40-year-olds. Wearable devices and much better testing should lower our lifestyle risks and extend our lives. AI has reduced the time required to develop new medicines at lower cost, but given pharmaceutical companies' profit incentives, those benefits may not extend much past the company itself.

AI promises diagnosis and remote access for the 6.5 billion people with inadequate health care. Most disruptions are from without, not within. The gatekeepers in the industry, drug companies, doctors' associations, and government regulations mean innovation is painfully slow. Releases by Google in 2024, and Microsoft in 2025 demonstrate we are on a path to these systems getting almost error-free in the next 5-10 years. It will be a massive weight off the shoulders of all health systems around the world. When are we likely to see it? Dementia diagnosis? 2 years? Machine learning for X-rays was developed in 2003, but it remains relatively uncommon in 2024. The development and release of mRNA vaccines during the COVID-19 pandemic demonstrated that vaccine development can be achieved in a matter of months, rather than the typical 12 years.

We should be optimistic about our ability to detect and reduce colon, prostate, and breast cancers by 90% within three years. We possess that technology today, but it's still in clinical trials. Will that happen? Yes, but don't be surprised if it take at least three times longer to be in wide use. Companies and governments prioritise avoiding mistakes over achieving success. In 2014, appetite hormones such as insulin and GLP-1-like compounds were targeted to address metabolic syndrome, weight loss, and type 2 diabetes. Yet, it has taken a decade for semaglutide and several other similar medications to be released, gain market acceptance, and help individuals lose 15% of their body weight.

Similarly, don't anticipate having a drug to extend your life expectancy to 100 years anytime soon, unless you are one of the billionaires.

We will see a rise in surgical robots and brain interface solutions. Robots are used in robot-assisted surgery for enhanced precision and control. They provide better visualisation and minimal invasive surgery and are currently still based on working from the outside in. Miniaturisation (and microbots) changes that paradigm to work from the inside out, but adoption will be slow. Developers of brain interfaces have a huge market of disabled people providing mobility, speech, hearing and even sight, and the cost of devices is more than offset by the lifetime care of disability. Within a decade, we will have multiple solutions for the wealthy, and within 20 years for all.

16.2 End of Toil & Drudgery

We are facing the end of working as we know it today. The dirty, the dangerous and the drudgery will be consigned to history, just as horses for work were lost to steam engines. For the first time since the Industrial Revolution people won't be enslaved to toil. Transitions take time. They start slow but ramp up fast. AI for service jobs/white-collar or knowledge workers will augment or automate tasks. Depending on the industry, AI will provide an economic boost to the services industry. Some predict 140% increases in productivity by 2030. In sectors such as education, health and government, professions will be augmented and automated with AI. Sectors including helpdesk, IT support and financial services are at risk of job losses of up to 80%.

We have seen the rapid rise of industrial robots, and they will continue to increase. Within the next five years, humanoid robots will start doing the tasks humans do in blue-collar sectors. Most robotic innovators are focused on factories, firstly their own, but some are focused on consumer uses. In most factories, there are multiple tasks such as handling, stacking and moving components around. Tasks will be pushed to humanoid robots, and as they learn new and multiple tasks, they may be able to do everything in a job. And because those robots operate in spaces designed for humans, humans can step in and do those tasks when needed for training or a shortage of robots.

The disruption and consequences will be as large and consequential as the disruption in food and agriculture. Artificial Intelligence is already augmenting or displacing humans doing work. The change has started. We see changes already but those changes will accelerate over the next five years. Complete changes may take 2 decades. The disruption is likely to roll out in discernible phases as follows:

- AI is augmenting and/or automating service jobs (white-collar/ knowledge workers) today. As LLMs improve, businesses will move from trials to full deployment. We see it in in sectors such as software development (mostly augmentation) and helpdesks (mostly automation). As AI improves, deployment will scale. AI is as dumb and slow as it will ever be.
- Industrial robots will continue to work in major factories and distribution centres, with AI improving steadily.
- AI in transportation will move from 60 cities to all cities in all Western countries over the next five years. By 2030, personal public transport will move from merely replacing taxi and Uber drivers to causing a reduction in cars. The first step will be from 2 vehicles to 1 per household. Associated with this deployment will be the collapse of auto manufacturers that failed to transition from internal combustion engines to autonomous EVs.
- Humanoid robots will be introduced in 2025/2027 to factories and find a ready use for doing tasks that don't warrant the investment in industrial robots.
- Small businesses that struggle to find employees will use humanoid robots to do tasks at a price that is not worthwhile to hire a human.
- By 2030, humanoid robots will start replacing most new jobs in business, hospitals, aged care, and homes and by 2035 replace existing jobs.
- Within a decade, AI and humanoid robotic labour will change geopolitics.
- By 2045, there will not no job that a machine cannot do.

There are 1 billion people globally who earn less than $10 daily, including the 60 million slaves. Humanoid robots will be cheaper than $2 per hour. What happens to those people, as they are unlikely to be in countries with a universal basic income?

A world with robots for every single person will need countries to provide a universal basic income (or a universal prosperity income). We measure GDP and use GDP per capita. What happens when the work and tasks get done without human labour cost, pollution, waste and zero marginal cost energy? How can we utilise those robots to handle tasks we don't like or can't currently afford to have a person perform? Replanting billions of trees and cleaning up the oceans of plastic?

Governments need to establish a roadmap today. AI will not stay still. Improvements will be at least 3 times more a year with no end in sight. Even the CEOs of major AI companies cannot forecast clarity of the future.

16.3 Food Disruption

It's very hard to get our heads around the fact that food disruption over the next decade will be bigger than the domestication of animals about 11,000 years ago. The population was then between 1 and 15 million, not 8.1 billion. Most of the 117 billion people who have ever lived on the planet have eaten domesticated animals. Before domestication, humans needed to hunt and kill prey. Now we make friends with them, but perversely, we kill and eat at least 1,000 animals each in our lifetime. Animals are very inefficient converters of energy. For every calorie of chicken, it takes nine calories to grow it. For dairy products, 40, and beef, 100. That is aside from the dumping of pesticides and antibiotics into all the food chains and fertilisers to pollute groundwater and water.

With the advances in genetics and multi-omics, for the first time ever, we can do with animals what we have done with plants for half a century. Cell culture and precision fermentation allow us to bypass the animal rearing and growing phases. All we will need is simple starches to create complex proteins. With technologies such as 3D printing, we will have the technology to produce food, the economics will be compelling, but will Tracey accept the chicken tender or the meat patty? She might for her cat or dog although there is evidence that people spend more on their pets than on their children. Substitution of a complex food such as a Wagyu steak will take another decade or two but the technology will get to a compelling price well before that luxury meat stage. The most rapid changes will be in processed food products. New entrants need to develop lab-grown food, get approval, and sell to a very limited number of business customers. Western consumers will be oblivious because the packaging on their food products will be the same. Economics is compelling for the manufacturers. Arguments that lab-grown food won't change rapidly ignore the fact that we have had new food sources in past centuries with almost no pushback. The rollout of lab-grown meat is likely to be in niche segments.

- Firstly, the pet food market, which is 20% of meat products.
- Egg production. The bakery industry is 50% of the egg market and do not need the shells that chickens produce.
- Chicken meat. Chicken tenders, which are at best 60% chicken meat, will be PF chicken. The industry will be 25% smaller by 2030 and a boutique industry by 2035.
- Palm oil. The industry will collapse. Firstly, with biofuel demand reduced by the move to EVs, and secondly, product for cosmetics and food products will be lab grown.
- Daily protein powder. Dairy protein will swamp the market by 2030.
- Fish protein.

What does this mean for the various food industries?

- The chicken industry will collapse by 50%.
- The dairy industry will be on its knees by 2035 and out of business by 2040.
- The beef meat industry will be 30% smaller by 2035.
- Fish protein spells the end of land-based fish farming by mid-2030s.
- Fishing in international waters will become uneconomical by 2035.

In Western countries, Tracey and her peers already consume ultra-processed foods. Ada and her children in Sub-Saharan Africa just need to feed their children enough. The unspoken challenge will be, what will all that agricultural land become? What will happen to those farmers and corporations who farm? These bold predictions seem implausible as we head down to the supermarket or the local farmers' market to buy fresh fruit and meat. Yet for 30% of Americans who never cook and the 80% who buy their food in a box or foil, there will be no change. For Tracey, fresh vegetables will continue to be grown on intensive farms, as they have been for the past 50 years. Consolidation among a few growers will continue, as it has done over the past few decades.

For Ada, she will continue to buy their vegetables at the local village market, and the change they will see will be cheaper protein.

16.4 Regreening of The Planet

The coal, gas and oil industry will leave scars on the planet for decades to come. Much is made of the space needed for renewable energy farms and the mining required for metals for the renewable energy transition, but that pales in comparison to the land and waste from the fossil fuel industry. ExxonMobil claimed no long-term damage from the Valdez oil spill in Prince William Sound in 1989 that fouled 2,200km of coastline, killing wildlife and ruining livelihoods. Three decades on, the oil is now buried, but if exposed, it is still as toxic as 35 years ago. Guillemots, herring and other species have gone. As Lydon [278] says, the disaster is one part hopefulness about nature's resiliency, one part outrage at the lingering damage, and a measure of urgency about our need to move toward a clean-energy future. At the BP Horizon oil spill in 2010, 14 years on, the spill that caused widespread destruction on the Louisiana coastline had oil contamination 10 times higher than pre-spill levels, destroyed coral reefs, cost 22,000 jobs, and BP paid just $20 billion. The long-term effects are expected to last four decades.

It is distressing that the millions of gas wells, oil wells and coal mines currently abandoned or in production will make Valdez and Horizon seem trivial in comparison. That legacy will haunt our children and grandchildren. A trillion-dollar shortfall exists for the remediation of mines and wells. Renewable energy technologies do require mines, but mining is 580 times less intensive than current fossil fuel mining. Fossil fuel mining could stop now, yet countries still approve new mines and oil and gas wells. Certainly, the transition to renewable energy will require the use of fossil fuels to generate the power needed to develop the technology for the transition. Once changed, we can begin the process of fixing up the mess left behind. Climate change is at its most critical era in 2040, as the methane we emit today will be at its worst. We have twenty years before the absolute end of the fossil fuel industry, so how do we minimise the additional emissions and pollution? The more we emit now, the bigger the task removing methane and CO_2 in the future. We need to electrify everything now and as fast as we can.

Fortunately, disruptions to food and agriculture will help remove methane and CO_2. Naturally. Without carbon capture or direct air capture technologies, which the fossil fuel era promotes. Regeneration of agricultural land will be the most effective and cheapest solution we have for regreening our planet. It has worked for billions of years! Free! Powered by the Sun. It's the most optimistic scenario we have in this world of bleakness and destruction by the fossil fuel industry in cahoots with the financiers and the politicians. Moreover, such regreening might seem like an enormous task, but in a decade, we will be able to use humanoid robots to do some of the manual work for less than $0.50 per hour, at 1/10th the cost of using human labour.

As food disruption shoulders its way into the food industry, the biggest change will be in the Western world. It won't be the small farmers but those few very large corporations that provide 70% of the food. We will see the end of large-scale corn and soybean farming for animal feed. Furthermore, the electrification of the vehicle fleet will slash corn and sugar cane farming for ethanol. With the 12 million hectares of land released from growing ethanol as transport electrifies, plus the millions of hectares of animal fodder, the sheer huge area of land released from cultivation is mind-boggling. At least 70% of the world's cropland will be retired. Additionally, in countries such as Brazil, Argentina and Australia, low-intensity grazing will end, and that will release 90% of the land used for grazing. Turn that into a vigorous forest, and we begin to reduce CO_2 in the atmosphere. Slowly, but steadily. It will take decades because we have been polluting for nearly a century.

16.5 Practical Steps for Transition

Education is key to this disruption.

- **Energy:** We have to counter the fossil fuel industry and focus on those 180 companies with their 1800 C-Suite and 20,000 politicians.
- **Health:** Don't let the entrenched drug companies play loose and fast. Doctors and physicians will not want to see their power diluted. Drug companies will want to maximise their return on investment.
- **Food**: Agriculture is a potent, powerful lobbying force in the Western world. That fanciful view that there are millions of farmers is false - just

a few hundred farmers control the Western food supply and vast land areas. The 600 million small farmers do not have a voice.

- **Labour:** Policies and support structures will need to be in place as AI replaces humans in the service industry and humanoid robots transform the blue-collar workforce. Forms of universal income need to be put in place today.
- **Politics:** Policy makers and politicians need to understand the changes coming and start the discussion and process today. It is inevitable that a large number of jobs will be supplanted by AI or AI-enabled transport and machines. We need to have societal support systems in place now.

Overall Conclusions:

- AI provides massive change across multiple disciplines – bringing both optimism and hard political questions.
- Healthcare advances – faster diagnosis, cheaper drugs, and a possible end to the worst of cancer within two decades.
- The end of toil and drudgery – how AI and robotics roll out in phases across the next two decades.
- Food disruption and the regreening of the planet – what happens to the land we free up?
- The 1,800 C-suite executives and 20,000 politicians who hold disproportionate power over our destiny.
- Practical steps for the transition across energy, health, food, labour, and politics – and what your part can be.

---

17 Dedication

To my long-suffering partner Judy, as I talk about incessantly. To those colleagues and friends who have worked with me over decades. To my children and grandchildren, for whom I wish to leave the world a more equitable place, a place where we live in harmony with nature.

18 Limits on Information

Change may be rapid. Information in this book may be dated. Hemingway in 1929. "How did you go bankrupt? Slowly and then suddenly." Innovation can take years to go from lab to full commercial deployment. Globally, R&D is over $2.2 trillion, and ranges from minimal to 6% in Israel.[279] I've been exposed to thousands of start-ups in my career, and it's this tsunami of great ideas. Most startups fail to see the light of day, while others change the world. A few get media attention, often because companies need money, but most are quietly introduced. That's partly why I believe there is a storm of optimism that will change our society. Many face barriers not just technology but also regulatory approval. Ideas to product can take 10 or 20 years - especially in health care. Some appear almost instant. ChatGPT AI was released in 2022 and became the fastest deployed software, but the birth of AI was in the 1950s!

Hang onto your hats. There are thousands of good things happening in clean energy, transportation, Artificial Intelligence, robotics, food and health care and will flood our lives over the next 2 decades. Understand that it is seldom just one breakthrough. The combination of multiple technologies delivers change bigger than just the sum of its parts. There are risks. We need to be very clear about what they are and we need to take steps to avoid them. We cannot rely on corporations. They are legally obliged to their shareholders to do the best by them. Many governments are too compromised to those corporations to consider their citizens. Action is needed.

The only thing necessary for the triumph of evil is for good men to do nothing. Attributed to Edmund Burke, including by John F Kennedy in a speech in 1961. Burke didn't say it, and its earliest form was by John Stuart Mill:

Bad men need nothing more to compass their ends than that good men should look on and do nothing. John Stuart Mill 1867:

19 Glossary

We use TLA's - three letter acronyms (WTF) or even FLAs (four letter ones), so here are some TLAs, FLAs and phrases used. throughout this book.

19.1.1 Energy Glossary

M = million
Bn = billion or 1,000 times more than a million
t= 1,000 billion
J=joule. A measure of energy. A joule of energy applied for 1 second is a watt. Therefore a Wh is 3,600 joules.
W= watt. A more commonly used form of energy measure on a day to day basis. An incandescent light bulb might be 60W
Wh = watt-hour. How much energy for an hour. A 60W light bulb would use about 60Wh if on for 1 hour. That energy is converted into light energy, and heat energy.
kW = 1,000 Watts

MW = 1m Watts.
Sun = a huge free nuclear fusion and fission energy source, that has lasted 6 billon years and due to last another 6b years
EEI = Earth energy imbalance. The Earth (more correctly Ocean) receives about 1370 watts per square meter (W/m2) in the tropics. All but 1 to 2 is irradiated back out.
RE = Renewable Energy
ICE = internal combustion engine
EV = Electric vehicle
BEV = Battery Electric Vehicle
Hybrid = an electric vehicle with an internal combustion engine
PHEV = A hybrid that can be plugged in to charge the battery
FCEV = a hybrid vehicle with an electric motor using a hydrogen fuel cell to charge the battery.
IPCC = Intergovernmental Panel on Climate Change
Moore's Law = the number of transistors on a chip doubles every 2 years (1965).
Swanson's Law = states solar module prices drop 20% for every doubling of installed capacity. Costs go down 75% per decade.
Wrights Law = As production increases, costs decline at a consistent rate as the cumulative (Wright 1936), sometimes referred to as learning curve effect. Predicts learning rate of 20-25% .
A climate denier = someone who ignores data and denies climate change
A climate delayer = someone who accepts climate change but takes not action
A nuclear advocate = A Climate delayer who proposes more nuclear in 15 years and keep burning fossil fuel
A hydrogen advocate = a climate delayer with hopium and little or zero science understanding
Malthusianism. A 19th century myth that there is a population bomb and thinks eugenics is a solution, as Germany did in the 1930's, or supports the UN Fund for Population Activities (UNFPA).
Doomster = sees climate change and thinks we cannot change the progression of climate change impacts and escapes to Bali to do regenerative agriculture
LCOE = Levelised cost of electricity. A measure to compare different technologies under a similar framework.

19.1.2 AI Glossary

- **Generative AI** refers to a type of artificial intelligence that can create new content, such as text, images, audio, and even code, in response to user prompts, and that can be by voice or text prompt such as CoPilot, ChatGPT, Claude and a whole raft of others. Computers talking to computers (using APIs) is also probably using generative AI.
- **Traditional or narrow AI** might be the AI driving Netflix recommendations or voice assistants such as Siri or Alexa.
- **Agentic AI** are neural networks used in autonomous driving or in AI Automation .
- **Machine Learning** (ML) is one of the foundational branches of AI. The goal of AI is to mimic human intelligence in general whereas ML is a subset of AI and is the science of developing algorithms and statistical models that computer systems use to perform complex tasks without explicit instructions. Think of the personalised recommendations on streaming services like Netflix or the content that appears on your social media feeds—all thanks to ML. That distinction will be lost as LLMs are already incorporating ML into their capabilities.
- **Transformers** are another key concept in neural networks. They were defined in 2017 and learn context in sequential data. These expanded from the previous ML work. [280] The transformer architecture is at the heart of groundbreaking models like ChatGPT. They've been instrumental in OpenAI's cutting-edge large language models (LLM) and played a key role in DeepMind's AlphaStar.
- **BERT** or Bidirectional Encoder Representations from Transformers and other transformer models made 2018 "a watershed moment" for natural language processing (NLP) Dar reported in 2018 [281]. "*Quite a few experts have claimed that the release of BERT marks a new era in NLP,*" it added. Developed by Google, BERT delivered state-of-the-art scores on benchmarks for NLP. In 2019, Google announced that BERT powers the company's search engine. Google then released BERT as open-source software, spawning a family of follow-ons and setting off a race to build ever larger, more powerful LLMs. Meta created an enhanced version called RoBERTa, released as open-source code in July 2017. For training, it used "*an order of magnitude more data than BERT,*" the paper said, and leapt ahead on NLP leader boards.

- **Large Language Models** (LLM) are focused on language and use all the technologies such as ML, Transformers and BERT. LLMs are used in generative AI chatbots like ChatGPT, Google Gemini, Grok, Meta AI, DeepSeek, Claude, Llama, Le Chat and Bing Chat plus another 100 or so to produce responses that resemble those of human beings. Their human-like responses to questions are enabled by combining deep learning and natural language generation algorithms with a large text library. Huge volumes of data are used to train LLMs, which employ a transformer neural network architecture specifically designed for language processing. There are also what are termed large multimodal models (LMMs) that include video, images and voice. The rise of LMMs are already in widespread use as well as LLMs.
- **Training versus Inference** AI can be compared with human intelligence. Neural nets have to be trained on data. A baby is born with some intelligence passed on by biology (DNA, genes), but it learns to adapt to the world over time. That is training. Most adults learn from their experience. Learning to drive a motor car can be challenging. Once we have learnt, we drive without thinking, even if there might be 18 different tasks to do when we engage the gears and pull away from traffic lights. That "automatic" part is mostly inference – we take the training data and deploy it. So too with digital AI. The models are trained. The part is inference.
- **Training Data**. Companies developing AI have utilised all available information on the internet for training, some of which is copyrighted, and there are ongoing legal disputes over this use. The ChatGPT chatbot was trained on a massive body of text data comprising approximately 570GB of datasets, including web pages, books, and other sources, and was trained with over 175 billion parameters. ChatGPT4 can handle over 25,000 words of text. The stated cost of computing resources to develop ChatGPT 4 was over $100m. As we will explore further, the costs to train models have plummeted, with DeepSeek and Le Chat costing less than an estimated $20 million.
- **Inference** is when you actually use that neural network. The majority of the computational cost is in developing the neural net and determining its "weights". That "runtime" version can be transferred to a completely different computer. Modern computers and mobile

phones may now include chips that can do AI inference on the actual device, rather than sending the instructions to a data centre. Tesla has taken this approach with its autonomous driving and humanoid robots, spending billions to develop the model on AI supercomputers, but in the car or on the robot, inference is done with a low-powered specialised $1,000 computer. That computer gets its underlying AI model and weights but can be updated with new weights and models to continually improve. If a car is self-driving, it does not want to rely on wireless communication for time-critical decision-making.

- **Superintelligence or AGI.** The terms "artificial general intelligence" (AGI) and "superintelligence" are used to describe AI that surpasses human capabilities in areas such as emotion, self-awareness, functional consciousness, and perception. The words are often used interchangeably. See the section on Superintelligence.
- **GPT – Generative Pre-trained Transformers** concept was introduced by OpenAI in 2018 and trained on vast amounts of data and could do generative AI.
- **Mixture of Experts (MoE) architecture** divides a large AI model into specialised sub-networks, called experts, each handling a different part of the input space. A "router" network then directs input to the most appropriate expert(s). This approach enables the construction of more efficient and scalable models, particularly for complex tasks such as natural language processing, computer vision, autonomous vehicles and robotics.
- **API, or Application Programming Interface**, is a defined way that computer programs communicate with other computer programs. Humans communicate with computers in natural language, and we can type or write at a rate of about 100 words per minute. Computer programs talk to each other via an API at megabits per second. Just like talking in English, APIs do need a defined structure. APIs have enabled rapid development of new applications – companies do not need to reinvent the core neural networks and the training needed.
- **GUI** is a graphical user interface, or how a user interacts with the software. We are familiar with the Windows GUI, or the Apple interface. Today, most LLM use a "dumb" text interface. It might include a microphone so we can use spoken word. APIs allow other

applications to have their own GUIs while connecting at computer speed to the LLM or LMM. I would predict that the current GUI will be completely different within years as the technology matures.

- **MCP.** Model context protocol. A new "glue" that enables Agentic AI to talk to different applications. It extends the concept of APIs that enables different computer systems to communicate with each other, MCP has connectors to different LLMs and different applications.

20 List of Figures

21 Citations

[1] 'Amazon.com: Brighter: Optimism, Progress, and the Future of Environmentalism eBook : Dorr, Adam: Kindle Store'. Accessed: Aug. 15, 2024. [Online]. Available: https://www.amazon.com/Brighter-Optimism-Progress-Future-Environmentalism-ebook/dp/B0BNYC1GWY

[2] B. to you byABC R. N. byAntony Funnell, 'Reassessing the laws of war; and why progress isn't always a straight line', ABC listen. Accessed: Jun. 19, 2025. [Online]. Available: https://www.abc.net.au/listen/programs/futuretense/laws-of-war-landmines-progress-/105340248

[3] B. Flyvbjerg and D. Gardner, *How Big Things Get Done: The Surprising Factors That Determine the Fate of Every Project, from Home Renovations to Space Exploration and Everything in Between*. New York: Currency, 2023.

[4] T. Seba, 'Rethinking Transportation 2020-2030 The Disruption of Transportation and the Collapse of the Internal-Combustion Vehicle and Oil Industries', RethinkX. Accessed: Feb. 03, 2020. [Online]. Available: https://www.rethinkx.com/transportation-executive-summary

[5] 'The world of work is being transformed', Grattan Institute. Accessed: Jun. 22, 2025. [Online]. Available: https://grattan.edu.au/news/the-world-of-work-is-being-transformed/

[6] 'Introduction To The ARK Innovation ETF (ARKK)', Ark Invest. Accessed: Feb. 21, 2025. [Online]. Available: https://www.ark-funds.com/articles/etf/Introduction To The ARK Innovation ETF (ARKK)

[7] 'Neural network (machine learning)', *Wikipedia*. Jun. 23, 2025. Accessed: Jun. 25, 2025. [Online]. Available: https://en.wikipedia.org/w/index.php?title=Neural_network_(machine_learning)&oldid=1296997155

[8] mkaczmarski, 'Liebreich: Generative AI – The Power and the Glory', BloombergNEF. Accessed: Jun. 24, 2025. [Online]. Available: https://about.bnef.com/insights/clean-energy/liebreich-generative-ai-the-power-and-the-glory/

[9] 'All Large Language Models Directory - All LLMs'. Accessed: May 07, 2024. [Online]. Available: https://llmmodels.org/

[10] 'Personalization Maturity - The Path to Hyperpersonalization | Adobe Experience Cloud'. Accessed: Jun. 17, 2025. [Online]. Available: https://experienceleague.adobe.com/en/docs/events/adobe-customer-success-webinar-recordings/2024/hyperpersonalization

[11] N. Kos'myna, 'Your Brain on ChatGPT: Accumulation of Cognitive Debt when Using an AI Assistant for Essay Writing Task', MIT Media Lab. Accessed: Jun. 19, 2025. [Online]. Available: https://www.media.mit.edu/publications/your-brain-on-chatgpt/
[12] C. Wang, 'Publishers facing existential threat from AI, Cloudflare CEO says', Axios. Accessed: Jun. 30, 2025. [Online]. Available: https://www.axios.com/2025/06/19/ai-search-traffic-publishers
[13] 'Universal Emotions | What are Emotions? | Paul Ekman Group'. Accessed: Jun. 27, 2025. [Online]. Available: https://www.paulekman.com/universal-emotions/
[14] N. Bostrom, *Superintelligence: Paths, Dangers, Strategies*. Grand Haven, MI: Audible Studios on Brilliance Audio, 2014.
[15] 'Why China's AI breakthroughs should come as no surprise', World Economic Forum. Accessed: Jul. 07, 2025. [Online]. Available: https://www.weforum.org/stories/2025/06/china-ai-breakthroughs-no-surprise/
[16] 'Global AI Power Rankings: Stanford HAI Tool Ranks 36 Countries in AI | Stanford HAI'. Accessed: Jul. 07, 2025. [Online]. Available: https://hai.stanford.edu/news/global-ai-power-rankings-stanford-hai-tool-ranks-36-countries-in-ai
[17] C. Wood, 'Big Ideas 2024 | ARK Invest', Ark Invest Big Ideas. Accessed: Apr. 27, 2024. [Online]. Available: https://ark-invest.com/big-ideas-2024
[18] 'Big Ideas 2023: Robotics And 3D Printing', Ark Invest. Accessed: Jun. 29, 2024. [Online]. Available: https://www.ark-invest.com/big-ideas-2023/3d-printing-and-robotics
[19] 'Elon Musk selects Memphis for "world's largest" supercomputer site | Fortune'. Accessed: Jun. 28, 2024. [Online]. Available: https://fortune.com/2024/06/18/elon-musk-xai-memphis-supercomputer-site/
[20] M. Zuhair, 'Intel Says It Won't Compete With NVIDIA In AI Market, Shifts Focus Towards Bringing Cost-Effective AI Solutions With Gaudi 3', Wccftech. Accessed: Jul. 09, 2025. [Online]. Available: https://wccftech.com/intel-gives-up-ai-race-nvidia-shifts-focus-towards-cost-effective-ai-solutions-gaudi-3/
[21] 'AI Data Center Build Advances at Full Speed: Five Things to Know | BloombergNEF'. Accessed: May 30, 2026. [Online]. Available: https://about.bnef.com/insights/commodities/ai-data-center-build-advances-at-full-speed-five-things-to-know/
[22] 'The cost of compute power: A $7 trillion race | McKinsey'. Accessed: Jun. 26, 2025. [Online]. Available: https://www.mckinsey.com/industries/technology-media-and-

telecommunications/our-insights/the-cost-of-compute-a-7-trillion-dollar-race-to-scale-data-centers
[23] D. S. H. your say, 'Nvidia and Elon Musk's xAI join $30bn AI data center fund from Microsoft and BlackRock'. Accessed: Jun. 26, 2025. [Online]. Available: https://www.datacenterdynamics.com/en/news/nvidia-and-elon-musks-xai-join-30bn-ai-data-center-fund-from-microsoft-and-blackstone/
[24] 'Liebreich: Generative AI – The Power and the Glory', BloombergNEF. Accessed: Jan. 30, 2025. [Online]. Available: https://about.bnef.com/blog/liebreich-generative-ai-the-power-and-the-glory/
[25] 'The Dawn of Stellar Energy: Announcing RethinkX's "Understanding Stellar Energy" Report'. Accessed: Jun. 08, 2025. [Online]. Available: https://www.rethinkx.com/blog/rethinkx/understanding-stellar-energy-is-here
[26] 'This Is the Next Big Power-Grid Investment - Altimetry'. Accessed: Jan. 27, 2025. [Online]. Available: https://altimetry.com/articles/this-is-the-next-big-power-grid-investment
[27] A. Staff, 'Amazon investing AU$20 billion to expand data center infrastructure in Australia and strengthen the nation's AI future'. Accessed: Jun. 20, 2025. [Online]. Available: https://www.aboutamazon.com/news/aws/amazon-data-center-investment-in-australia
[28] 'Eliezer Yudkowsky', *Wikipedia*. Jun. 01, 2025. Accessed: Jun. 22, 2025. [Online]. Available: https://en.wikipedia.org/w/index.php?title=Eliezer_Yudkowsky&oldid=1293428407
[29] 'Elon Musk says there's a 10% to 20% chance that AI "goes bad," even while he raises billions for his own startup xAI | Fortune'. Accessed: Jun. 22, 2025. [Online]. Available: https://fortune.com/2024/10/30/elon-musk-ai-could-go-bad-existential-threat-xai-fundraising/
[30] M. Field, 'Apathy and hyperbole cloud the real risks of AI bioweapons', Bulletin of the Atomic Scientists. Accessed: Jun. 26, 2025. [Online]. Available: https://thebulletin.org/2024/09/apathy-and-hyperbole-cloud-the-real-risks-of-ai-bioweapons/
[31] D. Milmo and D. M. G. technology editor, '"Godfather of AI" shortens odds of the technology wiping out humanity over next 30 years', *The Guardian*, Dec. 27, 2024. Accessed: Jun. 22, 2025. [Online]. Available: https://www.theguardian.com/technology/2024/dec/27/godfather-of-ai-raises-odds-of-the-technology-wiping-out-humanity-over-next-30-years

[32] 'The Risks Associated with Artificial General Intelligence: A Systematic Review'. Accessed: Jul. 08, 2025. [Online]. Available: https://airisk.mit.edu/blog/the-risks-associated-with-artificial-general-intelligence-a-systematic-review
[33] 'The Future of Disruption | RethinkX'. Accessed: Jun. 29, 2024. [Online]. Available: https://www.rethinkx.com/disruption
[34] 'Disruptive Innovation Investment Strategy by ARK Invest'. Accessed: Jun. 29, 2024. [Online]. Available: https://www.ark-invest.com/strategy/ark-disruptive-innovation
[35] Y. Shao *et al.*, 'Future of Work with AI Agents: Auditing Automation and Augmentation Potential across the U.S. Workforce', Jun. 11, 2025, *arXiv*: arXiv:2506.06576. doi: 10.48550/arXiv.2506.06576.
[36] 'AI Will Transform the Global Economy. Let's Make Sure It Benefits Humanity.', IMF. Accessed: Jun. 28, 2024. [Online]. Available: https://www.imf.org/en/Blogs/Articles/2024/01/14/ai-will-transform-the-global-economy-lets-make-sure-it-benefits-humanity
[37] 'Public sector employment and earnings, 2023-24 financial year | Australian Bureau of Statistics'. Accessed: Jun. 23, 2025. [Online]. Available: https://www.abs.gov.au/statistics/labour/employment-and-unemployment/public-sector-employment-and-earnings/latest-release
[38] 'Transfer fees for money sent home by international migrants were nearly as high as US foreign aid in 2023', Our World in Data. Accessed: Jun. 30, 2025. [Online]. Available: https://ourworldindata.org/data-insights/transfer-fees-for-money-sent-home-by-international-migrants-were-nearly-as-high-as-us-foreign-aid-in-2023
[39] 'BIG IDEAS 2025'. Accessed: Feb. 15, 2025. [Online]. Available: https://www.ark-invest.com/big-ideas-2025
[40] 'Wall Street Job Losses May Top 200,000 as AI Replaces Roles'. Accessed: Jan. 23, 2025. [Online]. Available: https://finance.yahoo.com/news/wall-street-job-losses-may-071500049.html
[41] 'OpenAI Codex'. Accessed: Jan. 27, 2025. [Online]. Available: https://openai.com/index/openai-codex/
[42] 'Tabnine: The AI coding assistant that you control', Tabnine. Accessed: Aug. 22, 2024. [Online]. Available: https://www.tabnine.com/blog/
[43] 'Software Development Job Postings on Indeed in the United States'. Accessed: Feb. 15, 2025. [Online]. Available: https://fred.stlouisfed.org/series/IHLIDXUSTPSOFTDEVE
[44] C. Lu, C. Lu, R. T. Lange, J. Foerster, J. Clune, and D. Ha, 'The AI Scientist: Towards Fully Automated Open-Ended Scientific Discovery', Aug. 15, 2024, *arXiv*: arXiv:2408.06292. doi: 10.48550/arXiv.2408.06292.

[45] 'Microsoft Build 2025', Microsoft Build 2025. Accessed: May 21, 2025. [Online]. Available: https://news.microsoft.com/build-2025/

[46] 'CEO of BT Warns AI May Lead to More Job Cuts at the Firm - Business I...', archive.md. Accessed: Jun. 24, 2025. [Online]. Available: https://archive.md/U1axs

[47] R. D. 2 min read, 'Klarna's AI replaced 700 workers. It now wants some of them back to improve customer service', Yahoo Finance. Accessed: Jul. 09, 2025. [Online]. Available: https://www.independent.co.uk/news/business/klarna-ceo-sebastian-siemiatkowski-ai-job-cuts-hiring-b2755580.html

[48] 'Artificial intelligence in education | UNESCO'. Accessed: Jun. 28, 2024. [Online]. Available: https://www.unesco.org/en/digital-education/artificial-intelligence

[49] 'Top 10 AI Graphic Design Tools Review'. Accessed: Jun. 28, 2024. [Online]. Available: https://topten.ai/best-ai-graphic-design-tool-review/

[50] 'Australian researchers prove world-leading flood sensing technology | NSW SES'. Accessed: Jun. 17, 2025. [Online]. Available: https://www.ses.nsw.gov.au/news/australian-researchers-prove-world-leading-flood-sensing-technology

[51] 'City Brain', *Wikipedia*. Jul. 26, 2024. Accessed: Jun. 24, 2025. [Online]. Available: https://en.wikipedia.org/w/index.php?title=City_Brain&oldid=1236706149

[52] 'Hangzhou launches City Brain 3.0, advancing smart governance'. Accessed: Jun. 24, 2025. [Online]. Available: https://www.ehangzhou.gov.cn/2025-04/01/c_293162.htm

[53] P. Hannam and P. H. E. correspondent, 'Australia's productivity plunge: why is the Reserve Bank concerned and should we panic?', *The Guardian*, Jun. 08, 2023. Accessed: Jun. 12, 2025. [Online]. Available: https://www.theguardian.com/business/2023/jun/09/australias-productivity-plunge-why-is-the-reserve-bank-concerned-and-should-we-panic

[54] 'Why Productivity Matters | Speeches', Reserve Bank of Australia. Accessed: Jun. 12, 2025. [Online]. Available: https://www.rba.gov.au/speeches/2025/sp-so-2025-02-27.html

[55] 'RethinkX Framework — RethinkX'. Accessed: May 06, 2024. [Online]. Available: https://www.rethinkx.com/framework

[56] 'This time, we are the horses: the disruption of labor by humanoid robots'. Accessed: May 06, 2024. [Online]. Available: https://www.rethinkx.com/blog/rethinkx/the-disruption-of-labour-by-humanoid-robots

[57] I. I. F. of Robotics, 'International Federation of Robotics', IFR International Federation of Robotics. Accessed: May 07, 2024. [Online]. Available: https://ifr.org/worldrobotics
[58] S. Sakama, Y. Tanaka, and A. Kamimura, 'Characteristics of Hydraulic and Electric Servo Motors', *Actuators*, vol. 11, no. 1, Art. no. 1, Jan. 2022, doi: 10.3390/act11010011.
[59] 'Electric Rod Actuators vs. Hydraulic Cylinders', Tolomatic. Accessed: Jan. 29, 2025. [Online]. Available: https://www.tolomatic.com/info-center/resource-details/electric-rod-actuators-vs-hydraulic-cylinders/
[60] 'Tesla Investor Relations'. Accessed: Feb. 16, 2025. [Online]. Available: https://ir.tesla.com/
[61] H. Mohapatra, 'Robotics x AI — a state of the union', Medium. Accessed: Jun. 15, 2025. [Online]. Available: https://medium.com/@MohapatraHemant/robotics-x-ai-a-state-of-the-union-7ce324ecb922
[62] D. Hardman, T. G. Thuruthel, and F. Iida, 'Multimodal information structuring with single-layer soft skins and high-density electrical impedance tomography', *Sci. Robot.*, vol. 10, no. 103, p. eadq2303, Jun. 2025, doi: 10.1126/scirobotics.adq2303.
[63] N. Zhang *et al.*, 'Biomimetic rigid-soft finger design for highly dexterous and adaptive robotic hands', *Sci. Adv.*, vol. 11, no. 17, p. eadu2018, Apr. 2025, doi: 10.1126/sciadv.adu2018.
[64] Jonas, Adam, 'Humanoids: The Humanoid 100: Mapping the Humanoid Robot Value Chain', *Morgan Stanley*, Feb. 2025, [Online]. Available: https://advisor.morganstanley.com/john.howard/documents/field/j/jo/john-howard/The_Humanoid_100_-_Mapping_the_Humanoid_Robot_Value_Chain.pdf
[65] 'China's EV giants are betting big on humanoid robots', MIT Technology Review. Accessed: Feb. 17, 2025. [Online]. Available: https://www.technologyreview.com/2025/02/14/1111920/chinas-electric-vehicle-giants-pivot-humanoid-robots/
[66] Cern Basher [@CernBasher], '@x_thors Thanks - they're already on the list - 8th one down (not ranked in any particular order). https://t.co/W914xGq2fl', Twitter. Accessed: Jun. 29, 2024. [Online]. Available: https://x.com/CernBasher/status/1788531898580369567
[67] T. R. R. Staff, 'Figure AI unveils BotQ high-volume humanoid manufacturing facility', The Robot Report. Accessed: Jun. 13, 2025. [Online]. Available: https://www.therobotreport.com/figure-ai-unveils-botq-high-volume-humanoid-manufacturing-facility/

[68] mattjackson_yaronfh8, ‘What Jobs Will AI Replace First?’, Bernard Marr. Accessed: Jun. 29, 2024. [Online]. Available: https://bernardmarr.com/what-jobs-will-ai-replace-first/
[69] ‘Will Automation Cause an Increase in the Unemployment Rate?’, Ark Invest. Accessed: Jun. 16, 2024. [Online]. Available: https://www.ark-invest.com/articles/analyst-research/will-automation-cause-an-increase-in-the-unemployment-rate/
[70] ‘These are the industries most likely to be taken over by robots’, World Economic Forum. Accessed: Jun. 16, 2025. [Online]. Available: https://www.weforum.org/stories/2019/04/these-are-the-industries-most-likely-to-be-taken-over-by-robots/
[71] ‘Harnessing the Tectonic Shifts in Global Manufacturing’, BCG Global. Accessed: Jul. 18, 2024. [Online]. Available: https://www.bcg.com/publications/2023/harnessing-tectonic-global-shift-in-manufacturing
[72] ‘Apple to move half of its supply chain from China to India, employ 5 lakh people in 3 years’, Firstpost. Accessed: Jul. 18, 2024. [Online]. Available: https://www.firstpost.com/tech/apple-to-move-half-of-its-supply-chain-from-china-to-india-employ-5-lakh-people-in-3-years-13758477.html
[73] B. Sawik *et al.*, ‘Robots for Elderly Care: Review, Multi-Criteria Optimization Model and Qualitative Case Study’, *Healthcare*, vol. 11, no. 9, p. 1286, Apr. 2023, doi: 10.3390/healthcare11091286.
[74] S. Wang, F. Wang, Z. Zhu, J. Wang, T. Tran, and Z. Du, ‘Artificial intelligence in education: A systematic literature review’, *Expert Syst. Appl.*, vol. 252, p. 124167, Oct. 2024, doi: 10.1016/j.eswa.2024.124167.
[75] ‘Key Work Health and Safety Statistics Australia 2024 | dataswa’. Accessed: Jun. 14, 2025. [Online]. Available: https://data.safeworkaustralia.gov.au/insights/key-whs-statistics-australia/latest-release
[76] ‘Slavery - Wikipedia’. Accessed: Apr. 26, 2024. [Online]. Available: https://en.wikipedia.org/wiki/Slavery
[77] ‘Infographic: The Hotspots of Modern Slavery’, Statista Daily Data. Accessed: Apr. 26, 2024. [Online]. Available: https://www.statista.com/chart/26321/countries-with-highest-estimated-prevalence-of-modern-slavery
[78] B. Wessling, ‘10 industries China is focusing on automating’, The Robot Report. Accessed: Jun. 29, 2024. [Online]. Available: https://www.therobotreport.com/china-plans-to-put-more-robots-into-these-10-industries/
[79] ‘Are we turning a corner on the humanoid robot age? | Perspectives | Macquarie Group’, Macquarie. Accessed: Jun. 29, 2024. [Online].

Available: https://www.macquarie.com/au/en/insights/are-we-turning-a-corner-on-the-humanoid-robot-age.html
[80] 'Elon Musk Backs Prediction of 1 Billion Humanoid Robots by 2040'. Accessed: Jun. 29, 2024. [Online]. Available: https://www.iotworldtoday.com/robotics/elon-musk-backs-prediction-of-1-billion-humanoid-robots-by-2040
[81] J. S. Toledano, B. D. Monedero, S. Flores - Ureba, and C. S. de Blas, 'The efficiency of urban public transport and its impact on environmental sustainability', *Sustain. Technol. Entrep.*, vol. 4, no. 2, p. 100097, May 2025, doi: 10.1016/j.stae.2025.100097.
[82] J. Göransson and H. Andersson, 'Factors that make public transport systems attractive: a review of travel preferences and travel mode choices', *Eur. Transp. Res. Rev.*, vol. 15, no. 1, p. 32, Sep. 2023, doi: 10.1186/s12544-023-00609-x.
[83] S. Rasca and N. Saeed, 'Exploring the factors influencing the use of public transport by commuters living in networks of small cities and towns', *Travel Behav. Soc.*, vol. 28, pp. 249–263, Jul. 2022, doi: 10.1016/j.tbs.2022.03.007.
[84] A. W. Team, 'New Insights for Scaling Laws in Autonomous Driving', Waymo. Accessed: Jun. 20, 2025. [Online]. Available: https://waymo.com/blog/2025/06/scaling-laws-in-autonomous-driving
[85] M. Baniodeh *et al.*, 'Scaling Laws of Motion Forecasting and Planning -- A Technical Report', arXiv.org. Accessed: Jun. 20, 2025. [Online]. Available: https://arxiv.org/abs/2506.08228v1
[86] 'Zoox: It's Not a Car', Zoox. Accessed: Jun. 24, 2025. [Online]. Available: https://zoox.com
[87] J. Shen, 'BYD says its $10,000 EV drives automatically without intervention · TechNode', TechNode. Accessed: Jul. 08, 2025. [Online]. Available: http://technode.com/2025/02/11/byd-says-its-10000-ev-can-drive-automatically-without-intervention/
[88] 'RACQ Vehicle Operating Costs 2023'. Accessed: Jul. 08, 2025. [Online]. Available: https://www.racq.com.au/about-us/news-and-media/news/2023/9/RACQ-Vehicle-Operating-Costs-2023
[89] 'How Many Uber Drivers Are There in 2024?' Accessed: Jun. 16, 2024. [Online]. Available: https://therideshareguy.com/how-many-uber-drivers-are-there/
[90] D. O. Memon, 'Pibot: How South Korea Developed An AI Robot That Can Fly Aircraft', Simple Flying. Accessed: Jun. 23, 2025. [Online]. Available: https://simpleflying.com/south-korea-pibot-robot-pilot-guide/

[91] 'Commercial Drones Market Research Report 2025, Trends Analysis And Market Growth'. Accessed: Jun. 21, 2025. [Online]. Available: https://www.thebusinessresearchcompany.com/report/commercial-drone-global-market-report
[92] 'Clouds or clear skies? Prospects for future air mobility'. Accessed: Jun. 21, 2025. [Online]. Available: https://www.mckinsey.com/industries/aerospace-and-defense/our-insights/future-air-mobility-blog/clouds-or-clear-skies-prospects-for-future-air-mobility
[93] D. Boffey, 'Killing machines: how Russia and Ukraine's race to perfect deadly pilotless drones could harm us all', *The Guardian*, Jun. 25, 2025. Accessed: Jun. 26, 2025. [Online]. Available: https://www.theguardian.com/world/2025/jun/25/ukraine-russia-autonomous-drones-ai
[94] Research and Markets 3 min read, 'Defense Drone Research Analysis Report 2025-2034: Counter-Drone Technologies See Surging Investments Amid Rising Threats, Autonomous and Swarming Drones at the Forefront of Warfare', Yahoo Finance. Accessed: Jun. 21, 2025. [Online]. Available: https://uk.finance.yahoo.com/news/defense-drone-research-analysis-report-145700555.html
[95] '(PDF) Global Consequences of Land Use'. Accessed: May 09, 2024. [Online]. Available: https://www.researchgate.net/publication/7704227_Global_Consequences_of_Land_Use
[96] H. Ritchie, 'If the world adopted a plant-based diet, we would reduce global agricultural land use from 4 to 1 billion hectares', *Our World Data*, Mar. 2021, Accessed: Jun. 20, 2025. [Online]. Available: https://ourworldindata.org/land-use-diets
[97] 'Land use statistics: England 2022', GOV.UK. Accessed: Jul. 09, 2024. [Online]. Available: https://www.gov.uk/government/statistics/land-use-in-england-2022/land-use-statistics-england-2022
[98] K. Pandey, 'India's built-up area grew by 2.5 million hectares in 17 years', Down To Earth. Accessed: Jul. 09, 2024. [Online]. Available: https://www.downtoearth.org.in/urbanisation/india-s-built-up-area-grew-by-2-5-million-hectares-in-17-years-95484
[99] K. Weisbrod, 'UN Report Says Humanity Has Altered 70 Percent of the Earth's Land, Putting the Planet on a "Crisis Footing"', Inside Climate News. Accessed: May 09, 2024. [Online]. Available: https://insideclimatenews.org/news/27042022/agriculture-land-report/
[100] 'Agricultural Subsidy Programs', Econlib. Accessed: Feb. 17, 2025. [Online]. Available:

https://www.econlib.org/library/Enc/AgriculturalSubsidyPrograms.html

[101] 'Government support to agriculture', OECD. Accessed: Feb. 17, 2025. [Online]. Available: https://www.oecd.org/en/topics/government-support-to-agriculture.html

[102] © Stanford University, Stanford, and California 94305, 'Do Government Subsidies Promote Productivity Growth in China?' Accessed: Feb. 23, 2025. [Online]. Available: https://sccei.fsi.stanford.edu/china-briefs/do-government-subsidies-promote-productivity-growth-china

[103] '(PDF) Land use changes during the past 300 years'. Accessed: May 22, 2024. [Online]. Available: https://www.researchgate.net/publication/242217587_Land_use_changes_during_the_past_300_years

[104] R. E. Green, S. J. Cornell, J. P. W. Scharlemann, and A. Balmford, 'Farming and the fate of wild nature', *Science*, vol. 307, no. 5709, pp. 550–555, Jan. 2005, doi: 10.1126/science.1106049.

[105] D. Tilman and C. Lehman, 'Human-caused environmental change: Impacts on plant diversity and evolution', *Proc. Natl. Acad. Sci.*, vol. 98, no. 10, pp. 5433–5440, May 2001, doi: 10.1073/pnas.091093198.

[106] F. Achard *et al.*, 'Determination of Deforestation Rates of the World's Humid Tropical Forests', *Science*, vol. 297, no. 5583, pp. 999–1002, Aug. 2002, doi: 10.1126/science.1070656.

[107] R. A. Houghton, 'Revised estimates of the annual net flux of carbon to the atmosphere from changes in land use and land management 1850–2000', *Tellus B*, vol. 55, no. 2, pp. 378–390, 2003, doi: 10.1034/j.1600-0889.2003.01450.x.

[108] A. J. Pitman, A. G. Slater, C. E. Desborough, and M. Zhao, 'Uncertainty in the simulation of runoff due to the parameterization of frozen soil moisture using the Global Soil Wetness Project methodology', *J. Geophys. Res.*, vol. 104, no. D14, pp. 16879–16888, Jul. 1999, doi: 10.1029/1999JD900261.

[109] I. R. Swingland *et al.*, 'The influence of land-use change and landscape dynamics on the climate system: relevance to climate-change policy beyond the radiative effect of greenhouse gases', *Philos. Trans. R. Soc. Lond. Ser. Math. Phys. Eng. Sci.*, vol. 360, no. 1797, pp. 1705–1719, Aug. 2002, doi: 10.1098/rsta.2002.1027.

[110] H. Ritchie and M. Roser, 'Water Use and Stress', *Our World Data*, Jul. 2018, Accessed: Feb. 02, 2025. [Online]. Available: https://ourworldindata.org/water-use-stress

[111] 'Tackling food loss and waste: A triple win opportunity'. Accessed: Feb. 18, 2025. [Online]. Available:

https://www.fao.org/newsroom/detail/FAO-UNEP-agriculture-environment-food-loss-waste-day-2022/en
[112] 'End Food Waste Australia - Food Waste Action» End Food Waste Australia'. Accessed: Feb. 18, 2025. [Online]. Available: https://endfoodwaste.com.au/
[113] 'Regenerative agriculture', *Wikipedia*. Apr. 12, 2024. Accessed: May 03, 2024. [Online]. Available: https://en.wikipedia.org/w/index.php?title=Regenerative_agriculture&oldid=1218633978
[114] P. Newton, N. Civita, L. Frankel-Goldwater, K. Bartel, and C. Johns, 'What Is Regenerative Agriculture? A Review of Scholar and Practitioner Definitions Based on Processes and Outcomes', *Front. Sustain. Food Syst.*, vol. 4, Oct. 2020, doi: 10.3389/fsufs.2020.577723.
[115] '5 benefits of regenerative agriculture – and 5 ways to scale it | World Economic Forum'. Accessed: Jul. 12, 2025. [Online]. Available: https://www.weforum.org/stories/2023/01/5-ways-to-scale-regenerative-agriculture-davos23/
[116] A. Brown, 'Electrifying the farm: "It could add a couple of hundred thousand to our bottom line"', *The Guardian*, Jun. 30, 2024. Accessed: Jul. 01, 2024. [Online]. Available: https://www.theguardian.com/australia-news/article/2024/jul/01/farms-switching-diesel-to-electric-australia
[117] 'Amazon.com: regenesis george monbiot'. Accessed: May 09, 2024. [Online]. Available: https://www.amazon.com/s/?ie=UTF8&keywords=regenesis+george+monbiot&index=aps&tag=hydglogoo-20&ref=pd_sl_rxlyrxxfk_e&adgrpid=155823215436&hvpone=&hvptwo=&hvadid=681378812611&hvpos=&hvnetw=g&hvrand=992664929843108399&hvqmt=e&hvdev=c&hvdvcmdl=&hvlocint=&hvlocphy=9072531&hvtargid=kwd-1640294451782&hydadcr=28017_14525551&gad_source=1
[118] M. N. Hayek and R. D. Garrett, 'Nationwide shift to grass-fed beef requires larger cattle population', *Environ. Res. Lett.*, vol. 13, no. 8, p. 084005, Jul. 2018, doi: 10.1088/1748-9326/aad401.
[119] 'Greenhouse gas emissions per 100 grams of protein', Our World in Data. Accessed: May 22, 2024. [Online]. Available: https://ourworldindata.org/grapher/ghg-per-protein-poore
[120] 'Organic beef lets the system down | Nature Food'. Accessed: May 22, 2024. [Online]. Available: https://www.nature.com/articles/s43016-020-0086-x
[121] T. D. Searchinger, S. Wirsenius, T. Beringer, and P. Dumas, 'Assessing the efficiency of changes in land use for mitigating climate change',

Nature, vol. 564, no. 7735, pp. 249–253, Dec. 2018, doi: 10.1038/s41586-018-0757-z.
[122] 'Biofuels - Energy System', IEA. Accessed: Jun. 29, 2024. [Online]. Available: https://www.iea.org/energy-system/low-emission-fuels/biofuels
[123] Martin, 'UN Report: Nature's Dangerous Decline "Unprecedented"; Species Extinction Rates "Accelerating"', United Nations Sustainable Development. Accessed: Jul. 07, 2024. [Online]. Available: https://www.un.org/sustainabledevelopment/blog/2019/05/nature-decline-unprecedented-report/
[124] 'Past and future decline and extinction of species | Royal Society'. Accessed: Jul. 07, 2024. [Online]. Available: https://royalsociety.org/news-resources/projects/biodiversity/decline-and-extinction/
[125] J. Allan, J. Watson, K. Jones, P. Negret, R. Fuller, and S. Maxwell, 'One-third of the world's nature reserves are under threat from humans', The Conversation. Accessed: Jul. 07, 2024. [Online]. Available: http://theconversation.com/one-third-of-the-worlds-nature-reserves-are-under-threat-from-humans-96721
[126] '30 by 30', *Wikipedia*. Jun. 30, 2024. Accessed: Jul. 07, 2024. [Online]. Available: https://en.wikipedia.org/w/index.php?title=30_by_30&oldid=1231835430
[127] 'Wetlands to be restored, protected for our community', OurSC. Accessed: Jul. 07, 2024. [Online]. Available: https://oursc.com.au/environment/wetlands-to-be-restored-protected-for-our-community
[128] UN Environment, Ed., 'Future Developments Without Targeted Policies', in *Global Environment Outlook – GEO-6: Healthy Planet, Healthy People*, Cambridge: Cambridge University Press, 2019, pp. 484–509. doi: 10.1017/9781108627146.027.
[129] V. Nechifor and M. Winning, 'Global Economic and Food Security Impacts of Demand-Driven Water Scarcity—Alternative Water Management Options for a Thirsty World', *Water*, vol. 10, no. 10, Art. no. 10, Oct. 2018, doi: 10.3390/w10101442.
[130] T. W. Counts, 'When will the world run out of water?' Accessed: Jul. 07, 2024. [Online]. Available: https://www.theworldcounts.com/challenges/planet-earth/state-of-the-planet/when-will-the-world-run-out-of-water
[131] R. Salvidge, 'England's rivers "under threat" as water extraction surges to record levels', *The Guardian*, Jun. 20, 2025. Accessed: Jun. 21, 2025. [Online]. Available:

https://www.theguardian.com/environment/2025/jun/20/england-rivers-water-extraction-record-levels
[132] 'Domestication Timeline | AMNH', American Museum of Natural History. Accessed: May 10, 2025. [Online]. Available: https://www.amnh.org/exhibitions/horse/domesticating-horses/domestication-timeline
[133] 'The State of Global Policy on Alternative Proteins - The Good Food Institute'. Accessed: Jun. 15, 2024. [Online]. Available: https://gfi.org/resource/alternative-proteins-state-of-global-policy/
[134] 'The Hacking of the American Mind: The Science Behind the Corporate Takeover of Our Bodies and Brains: Lustig, Robert H.: Amazon.com.au: Books'. Accessed: Jun. 10, 2024. [Online]. Available: https://www.amazon.com.au/Hacking-American-Mind-Corporate-Takeover/dp/1101982586
[135] H. Butler, *The Take-Out Diet: Science shows us how diets fail. Science and Business show us how diets work*. S.l.: Hugh Butler, 2014.
[136] 'The 15 Largest Seafood Suppliers In The World', Zippia. Accessed: Jun. 09, 2024. [Online]. Available: https://www.zippia.com/advice/largest-seafood-suppliers/
[137] S. K. Lowder, M. V. Sánchez, and R. Bertini, 'Which farms feed the world and has farmland become more concentrated?', *World Dev.*, vol. 142, p. 105455, Jun. 2021, doi: 10.1016/j.worlddev.2021.105455.
[138] 'Small family farmers produce a third of the world's food', Newsroom. Accessed: Jul. 02, 2024. [Online]. Available: https://www.fao.org/newsroom/detail/Small-family-farmers-produce-a-third-of-the-world-s-food/en
[139] J. Reedy *et al.*, 'Global Intakes of Total Protein and Sub-types; Findings from the 2015 Global Dietary Database (P10-050-19)', *Curr. Dev. Nutr.*, vol. 3, no. Suppl 1, p. nzz034.P10-050-19, Jun. 2019, doi: 10.1093/cdn/nzz034.P10-050-19.
[140] Stefan Wirsenius, 'Human use of land and organic materials: Modeling the turnover of biomass in the global food system'. Accessed: Jul. 03, 2024. [Online]. Available: https://www.researchgate.net/publication/294230838_Human_use_of_land_and_organic_materials_Modeling_the_turnover_of_biomass_in_the_global_food_system
[141] '17 of the Worst Foodborne Illness Outbreaks in U.S. History', Healthline. Accessed: Jun. 20, 2025. [Online]. Available: https://www.healthline.com/health/worst-foodborne-illness-outbreaks
[142] 'Food safety'. Accessed: Jul. 01, 2024. [Online]. Available: https://www.who.int/news-room/fact-sheets/detail/food-safety

[143] 'The Future of Food & Agriculture | RethinkX'. Accessed: May 13, 2024. [Online]. Available: https://www.rethinkx.com/food-and-agriculture
[144] 'State of the Industry: Fermentation for meat, seafood, eggs, dairy, and ingredients | GFI'. Accessed: May 11, 2025. [Online]. Available: https://gfi.org/resource/fermentation-meat-seafood-eggs-dairy-and-ingredients-state-of-the-industry/
[145] 'Fake Meat, What's The Beef? Ep166: Bruce Friedrich', Cleaning Up. Leadership in an age of climate change. Accessed: Jun. 15, 2024. [Online]. Available: https://www.cleaningup.live/fake-meat-whats-the-beef-ep166-bruce-friedrich/
[146] A. Horn-Muller, 'The climate case for mock meats is clear. But who can afford them?', Grist. Accessed: Jul. 02, 2024. [Online]. Available: https://grist.org/food-and-agriculture/the-climate-case-for-mock-meats-is-clear-but-who-can-afford-them/
[147] M. Valley, 'Magic Valley', Magic Valley. Accessed: Jun. 24, 2025. [Online]. Available: https://magicvalley.com.au/gallery
[148] 'Lab-grown meat has just been approved for consumption in Australia. What is it and how is it made? - ABC News'. Accessed: Jun. 24, 2025. [Online]. Available: https://www.abc.net.au/news/2025-06-24/lab-grown-meat-approved-for-sale-australia/105448604
[149] L. Turner, 'Agronomics Portco: Onego Bio's Pioneering Biotech Platform for Sustainable Egg Protein', Medium. Accessed: Jul. 03, 2024. [Online]. Available: https://medium.com/@laureturner_/agronomics-doubles-down-on-onego-bios-pioneering-biotech-platform-for-sustainable-egg-protein-ffb91ef3a7c9
[150] H. Ritchie, P. Rosado, and M. Roser, 'Meat and Dairy Production', *Our World Data*, Jan. 2024, Accessed: Jul. 03, 2024. [Online]. Available: https://ourworldindata.org/meat-production
[151] H. K. Lotze *et al.*, 'Global ensemble projections reveal trophic amplification of ocean biomass declines with climate change', *Proc. Natl. Acad. Sci.*, vol. 116, no. 26, pp. 12907–12912, Jun. 2019, doi: 10.1073/pnas.1900194116.
[152] 'Too warm to swim'. Accessed: May 12, 2024. [Online]. Available: https://www.stockholmresilience.org/research/research-news/2019-06-12-too-warm-to-swim.html
[153] 'More than a million salmon dumped after "unprecedented" mass mortality at Tasmanian fish farms | Tasmania | The Guardian'. Accessed: May 11, 2025. [Online]. Available: https://www.theguardian.com/australia-news/2025/mar/08/million-salmon-dumped-in-landfill-after-unprecedented-mass-death-of-at-tasmanian-fish-farms-ntwnfb

[154] 'The next wave: Alternative seafood solutions | McKinsey'. Accessed: May 13, 2024. [Online]. Available: https://www.mckinsey.com/industries/agriculture/our-insights/the-next-wave-alternative-seafood-solutions
[155] 'Israeli firm is the latest to print out a "real beef" lab-grown steak'. Accessed: Jun. 10, 2024. [Online]. Available: https://newatlas.com/science/meatech-3d-lab-grown-steak/
[156] 'Lab-grown Wagyu beef morsels could be coming to a meat market near you'. Accessed: Jun. 10, 2024. [Online]. Available: https://newatlas.com/science/lab-grown-waygu-omakase-beef-morsels/
[157] '"World's first" cultivated fish fillets served up in Israel', New Atlas. Accessed: May 12, 2024. [Online]. Available: https://newatlas.com/science/steakholder-foods-cultivated-fish-fillets-grouper/
[158] 'The quest to find healthy and cheap sweeteners'. Accessed: Jun. 10, 2024. [Online]. Available: https://www.bbc.com/news/business-67448897
[159] 'Innovation in precision fermentation for food ingredients'. Accessed: Jun. 10, 2024. [Online]. Available: https://www.tandfonline.com/doi/full/10.1080/10408398.2023.2166014
[160] 'Rethinking Food & Agriculture (2019) (English)'. Accessed: Jun. 11, 2024. [Online]. Available: https://www.rethinkx.com/publications/rethinkingfoodandagricultur e2019.en
[161] 'BioCraft's cultivated mouse meat reaches price parity with premium pet food', PetfoodIndustry. Accessed: Jun. 11, 2024. [Online]. Available: https://www.petfoodindustry.com/ingredient-issues/proteins/news/15676448/biocrafts-cultivated-mouse-meat-reaches-price-parity-with-premium-pet-food
[162] 'Estonian food tech startup Äio secures €6 million to turn wood into oils'. Accessed: Oct. 03, 2024. [Online]. Available: https://estonianworld.com/business/estonian-food-tech-startup-aio-secures-e6-million-for-demo-plant/
[163] 'Big brands invest in precision fermentation-derived palm oil'. Accessed: Jun. 11, 2024. [Online]. Available: https://www.ingredientsnetwork.com/big-brands-invest-in-precision-fermentation-news123560.html
[164] 'Precision Fermentation and the disruption of the palm oil Industry'. Accessed: Jun. 11, 2024. [Online]. Available: https://www.rethinkx.com/blog/precision-fermentation-and-the-disruption-of-the-palm-oil-industry

[165] M. Roser, 'How many animals get slaughtered every day?', *Our World Data*, Jan. 2024, Accessed: Jun. 11, 2024. [Online]. Available: https://ourworldindata.org/how-many-animals-get-slaughtered-every-day

[166] 'Agricultural subsidy', *Wikipedia*. May 19, 2024. Accessed: Jun. 11, 2024. [Online]. Available: https://en.wikipedia.org/w/index.php?title=Agricultural_subsidy&oldid=1224679837

[167] C. Li, Z. Sha, X. Sun, and Y. Jiao, 'The Effectiveness Assessment of Agricultural Subsidy Policies on Food Security: Evidence from China's Poverty-Stricken Villages', *Int. J. Environ. Res. Public. Health*, vol. 19, no. 21, p. 13797, Oct. 2022, doi: 10.3390/ijerph192113797.

[168] Ehrlich, Paul and Ehrlich, Anne, 'The Population Bomb', *Wikipedia*. May 19, 2024. Accessed: Jul. 02, 2024. [Online]. Available: https://en.wikipedia.org/w/index.php?title=The_Population_Bomb&oldid=1224583149

[169] Cor_admin, 'New Report to the Club of Rome: Sustainable Future – Overcoming the Barriers', Club of Rome. Accessed: Jul. 04, 2024. [Online]. Available: https://www.clubofrome.org/impact-hubs/rethinking-finance/new-report-to-the-club-of-rome-sustainable-future-overcoming-the-barriers-2/

[170] E. O. Wilson, *The Social Conquest of Earth*. New York: Liveright, 2012.

[171] 'Human Population Nearly Went Extinct 900,000 Years Ago, Genetic Study Suggests', GenomeWeb. Accessed: Jul. 04, 2024. [Online]. Available: https://www.genomeweb.com/sequencing/human-population-nearly-went-extinct-900000-years-ago-genetic-study-suggests

[172] 'Silurian hypothesis', *Wikipedia*. May 12, 2025. Accessed: May 16, 2025. [Online]. Available: https://en.wikipedia.org/w/index.php?title=Silurian_hypothesis&oldid=1290107806

[173] T. C. of Rome, 'Report: Global population could peak below 9 billion in 2050s'. Accessed: Jun. 17, 2025. [Online]. Available: https://phys.org/news/2023-03-global-population-peak-billion-2050s.html

[174] T. Daniels, 'Surprising study shows birth projections for 2026: only 8 out of every 100 babies will be born on three continents combined', ZENIT - English. Accessed: Jun. 12, 2026. [Online]. Available: https://zenit.org/2026/03/04/surprising-study-shows-birth-projections-for-2026-only-8-out-of-every-100-babies-will-be-born-on-three-continents-combined/

[175] M. Lu, 'Life Expectancy by Region (1950-2050F)', Visual Capitalist. Accessed: May 09, 2024. [Online]. Available: https://www.visualcapitalist.com/life-expectancy-by-region-1950-2050f/
[176] S. Dattani, F. Spooner, H. Ritchie, and M. Roser, 'Child and Infant Mortality', *Our World Data*, Dec. 2023, Accessed: May 12, 2024. [Online]. Available: https://ourworldindata.org/child-mortality
[177] M. Roser, 'Child mortality: an everyday tragedy of enormous scale that we can make progress against', *Our World Data*, Jul. 2021, Accessed: Jun. 13, 2026. [Online]. Available: https://ourworldindata.org/child-mortality-big-problem-in-brief
[178] J. Q. Blogstack, 'Are pro-natalists living on the same planet?', John Quiggin's Blogstack. Accessed: Jun. 19, 2025. [Online]. Available: https://johnquigginblog.substack.com/p/are-pro-natalists-living-on-the-same
[179] '*An Essay on the Principle of Population*', *Wikipedia*. Apr. 13, 2026. Accessed: Jun. 13, 2026. [Online]. Available: https://en.wikipedia.org/w/index.php?title=An_Essay_on_the_Principle_of_Population&oldid=1348581267
[180] 'A Scientist Says Humans Will Go Backwards in Time Within Just 5 Years', Popular Mechanics. Accessed: May 21, 2025. [Online]. Available: https://www.popularmechanics.com/science/a62990579/humans-backwards-in-time/
[181] Apur, 'Population density of Paris', Apur. Accessed: Jul. 04, 2024. [Online]. Available: https://www.apur.org/en/geo-data/population-density
[182] V. Capitalist, 'Mapped: North America Population Patterns by Density'. Accessed: Jun. 18, 2025. [Online]. Available: https://www.voronoiapp.com/demographics/Mapped-North-America-Population-Patterns-by-Density-765
[183] 'How we feed the world today - OECD'. Accessed: Jun. 20, 2024. [Online]. Available: https://www.oecd.org/agriculture/understanding-the-global-food-system/how-we-feed-the-world-today/
[184] 'Norman Borlaug - Wikipedia'. Accessed: Jul. 04, 2024. [Online]. Available: https://en.wikipedia.org/wiki/Norman_Borlaug
[185] 'Billionaire wealth surges by $2 trillion in 2024, three times faster than the year before, while the number of people living in poverty has barely changed since 1990', Oxfam International. Accessed: Jan. 23, 2025. [Online]. Available: https://www.oxfam.org/en/press-releases/billionaire-wealth-surges-2-trillion-2024-three-times-faster-year-while-number

[186] 'Rigged – How networks of powerful mates rip off everyday Australians'. Accessed: May 22, 2024. [Online]. Available: https://gameofmates.com/
[187] D. Neufeld, 'Visualizing the Pyramid of Global Wealth Distribution', Visual Capitalist. Accessed: Jul. 04, 2024. [Online]. Available: https://www.visualcapitalist.com/global-wealth-distribution/
[188] P. M. Forster *et al.*, 'Indicators of Global Climate Change 2024: annual update of key indicators of the state of the climate system and human influence', *Earth Syst. Sci. Data*, vol. 17, no. 6, pp. 2641–2680, Jun. 2025, doi: 10.5194/essd-17-2641-2025.
[189] Ritchie, Hannah, 'Vaccines have reduced infant mortality by 40% over last 50 years', Our World in Data. Accessed: Sep. 06, 2024. [Online]. Available: https://ourworldindata.org/data-insights
[190] N. May, 'I'm so sorry, I've got cancer: why do two of every five Australian men die early from preventable conditions?', *The Guardian*, Jul. 21, 2024. Accessed: Jul. 22, 2024. [Online]. Available: https://www.theguardian.com/australia-news/article/2024/jul/21/im-so-sorry-ive-got-cancer-why-do-two-of-every-five-australian-men-die-early-from-preventable-conditions
[191] R. N. Criner and M. K. Han, 'COPD Care in the 21st Century: A Public Health Priority', *Respir. Care*, vol. 63, no. 5, pp. 591–600, May 2018, doi: 10.4187/respcare.06276.
[192] D. S. P. Ph.D, 'What is Multiomics?', News-Medical. Accessed: Jul. 09, 2024. [Online]. Available: https://www.news-medical.net/life-sciences/What-is-Multiomics.aspx
[193] M. Babu and M. Snyder, 'Multi-Omics Profiling for Health', *Mol. Cell. Proteomics MCP*, vol. 22, no. 6, p. 100561, Apr. 2023, doi: 10.1016/j.mcpro.2023.100561.
[194] S. Fuchs and R. Milo, 'Are We Really Vastly Outnumbered? Revisiting the Ratio of Bacterial to Host Cells in Humans', *Cell*, Jan. 2016.
[195] P. Rao, 'Which Countries Have the Highest Infant Mortality Rates?', Visual Capitalist. Accessed: Jun. 16, 2024. [Online]. Available: https://www.visualcapitalist.com/countries-highest-infant-mortality-rates/
[196] 'Older Australians, Demographic profile', Australian Institute of Health and Welfare. Accessed: Jun. 16, 2024. [Online]. Available: https://www.aihw.gov.au/reports/older-people/older-australians/contents/demographic-profile
[197] S. Simmons-Duffin, '"Live free and die?" The sad state of U.S. life expectancy', *NPR*, Mar. 25, 2023. Accessed: Jun. 16, 2024. [Online]. Available: https://www.npr.org/sections/health-

shots/2023/03/25/1164819944/live-free-and-die-the-sad-state-of-u-s-life-expectancy

[198] A. Hamann, 'Aktuelles zur Adipositas (mit und ohne Diabetes)'. Accessed: May 07, 2024. [Online]. Available: https://www.researchgate.net/publication/318228076_Aktuelles_zur_Adipositas_mit_und_ohne_Diabetes

[199] C. Rampell, 'The World Is Fat', Economix Blog. Accessed: May 07, 2024. [Online]. Available: https://archive.nytimes.com/economix.blogs.nytimes.com/2010/09/23/the-world-is-fat/

[200] J. Ma *et al.*, 'The American Cancer Society 2035 challenge goal on cancer mortality reduction', *CA. Cancer J. Clin.*, vol. 69, no. 5, pp. 351–362, 2019, doi: 10.3322/caac.21564.

[201] 'Cancer Today'. Accessed: May 07, 2024. [Online]. Available: https://gco.iarc.who.int/today/

[202] 'ARK Invest | We Believe Innovation Is Key to Growth'. Accessed: Feb. 13, 2025. [Online]. Available: https://www.ark-invest.com/thank-you-big-ideas-2025?submissionGuid=20b59224-8bd0-4a63-a814-6bf57b80fd5f

[203] Y. Zhang and X. Yuan, 'Minimal residue disease detection in early-stage breast cancer: a review', *Mol. Biol. Rep.*, vol. 52, no. 1, p. 106, Jan. 2025, doi: 10.1007/s11033-024-10198-0.

[204] 'National Lung Cancer Screening Program to launch July 2025', Australian Medical Association. Accessed: Jun. 04, 2025. [Online]. Available: https://www.ama.com.au/ama-rounds/24-january-2025/articles/national-lung-cancer-screening-program-launch-july-2025

[205] '"Something's happened": Australians under 50 developing cancer at "alarming" rates', *ABC News*, Jul. 06, 2025. Accessed: Jul. 07, 2025. [Online]. Available: https://www.abc.net.au/news/2025-07-07/cancer-diagnosis-rates-under-50s-rising-causes-four-corners/105495620

[206] 'World Heart Report 2023: Full Report', World Heart Federation. Accessed: Jun. 15, 2024. [Online]. Available: https://world-heart-federation.org/resource/world-heart-report-2023/

[207] X. Shen *et al.*, 'Nonlinear dynamics of multi-omics profiles during human aging', *Nat. Aging*, pp. 1–16, Aug. 2024, doi: 10.1038/s43587-024-00692-2.

[208] 'Anti-ageing "supermodel granny" drug extends life in animal tests'. Accessed: Jul. 18, 2024. [Online]. Available: https://www.bbc.com/news/articles/cv2gr3x3xkno

[209] L. Cao, S. G. Lee, K. T. Lim, and H.-R. Kim, 'Potential Anti-Aging Substances Derived from Seaweeds', *Mar. Drugs*, vol. 18, no. 11, p. 564, Nov. 2020, doi: 10.3390/md18110564.

[210] E. Nichols *et al.*, 'Estimation of the global prevalence of dementia in 2019 and forecasted prevalence in 2050: an analysis for the Global Burden of Disease Study 2019', *Lancet Public Health*, vol. 7, no. 2, pp. e105–e125, Feb. 2022, doi: 10.1016/S2468-2667(21)00249-8.

[211] A. Javeed, A. L. Dallora, J. S. Berglund, A. Ali, L. Ali, and P. Anderberg, 'Machine Learning for Dementia Prediction: A Systematic Review and Future Research Directions', *J. Med. Syst.*, vol. 47, no. 1, p. 17, Feb. 2023, doi: 10.1007/s10916-023-01906-7.

[212] T. Conversation, 'The Roots of Dementia Trace Back All The Way to Childhood, Experts Say', ScienceAlert. Accessed: May 21, 2025. [Online]. Available: https://www.sciencealert.com/?p=161692

[213] P. Belluck, 'Dementia Cases in the U.S. Will Surge in the Coming Decades, Researchers Say', *The New York Times*, Jan. 13, 2025. Accessed: Jan. 28, 2025. [Online]. Available: https://www.nytimes.com/2025/01/13/health/dementia-cases-us.html

[214] 'Lifetime risk and projected burden of dementia | Nature Medicine'. Accessed: Jan. 28, 2025. [Online]. Available: https://www.nature.com/articles/s41591-024-03340-9

[215] 'An online multidomain lifestyle intervention to prevent cognitive decline in at-risk older adults: a randomized controlled trial | Nature Medicine'. Accessed: Feb. 02, 2025. [Online]. Available: https://www.nature.com/articles/s41591-024-03351-6

[216] 'Exercise Research', Wendy Suzuki. Accessed: May 28, 2024. [Online]. Available: https://www.wendysuzuki.com/exercise-research

[217] J. Snape, 'I ditched the gym and you can too – here are six ways to get fit without it', *The Guardian*, Jun. 15, 2025. Accessed: Jun. 16, 2025. [Online]. Available: https://www.theguardian.com/lifeandstyle/2025/jun/15/i-ditched-the-gym-and-you-can-too-here-are-six-ways-to-get-fit-without-it

[218] G. Livingston *et al.*, 'Dementia prevention, intervention, and care: 2020 report of the Lancet Commission', *The Lancet*, vol. 396, no. 10248, pp. 413–446, Aug. 2020, doi: 10.1016/S0140-6736(20)30367-6.

[219] 'Associations of early-life factors and risk of incident dementia: a cohort study of 247,841 participants - Ou - 2024 - Alzheimer's & Dementia - Wiley Online Library'. Accessed: May 21, 2025. [Online]. Available: https://alz-journals.onlinelibrary.wiley.com/doi/10.1002/alz.084318

[220] 'Artificial intelligence outperforms clinical tests at predicting progress of Alzheimer's disease | University of Cambridge'. Accessed: Jul. 18, 2024. [Online]. Available:

https://www.cam.ac.uk/research/news/artificial-intelligence-outperforms-clinical-tests-at-predicting-progress-of-alzheimers-disease
[221] 'AI techniques massively accelerate the search for Parkinson's disease treatments', News-Medical. Accessed: Jan. 24, 2025. [Online]. Available: https://www.news-medical.net/news/20240417/AI-techniques-massively-accelerate-the-search-for-Parkinsons-disease-treatments.aspx
[222] R. I. Horne *et al.*, 'Discovery of potent inhibitors of α-synuclein aggregation using structure-based iterative learning', *Nat. Chem. Biol.*, vol. 20, no. 5, pp. 634–645, May 2024, doi: 10.1038/s41589-024-01580-x.
[223] eClinicalMedicine, 'A new hope for patients with Huntington's disease?', *eClinicalMedicine*, vol. 88, p. 103612, Oct. 2025, doi: 10.1016/j.eclinm.2025.103612.
[224] G. Kolata, 'The Researcher Who Didn't Want to Know', *The New York Times*, Jun. 11, 2026. Accessed: Jun. 16, 2026. [Online]. Available: https://www.nytimes.com/2026/06/11/science/researcher-huntingtons-disease-wexler.html
[225] T. W. Counts, 'Chemical pollution'. Accessed: Jul. 10, 2024. [Online]. Available: https://www.theworldcounts.com/challenges/planet-earth/state-of-the-planet/chemical-pollution
[226] 'Toxic arsenic and lead found to be common in tampons', New Atlas. Accessed: Jul. 10, 2024. [Online]. Available: https://newatlas.com/health-wellbeing/tampons-toxic-metals/
[227] R. Naidu *et al.*, 'Chemical pollution: A growing peril and potential catastrophic risk to humanity', *Environ. Int.*, vol. 156, p. 106616, Nov. 2021, doi: 10.1016/j.envint.2021.106616.
[228] H. Ritchie, V. Samborska, and M. Roser, 'Plastic Pollution', *Our World Data*, Nov. 2023, Accessed: Jul. 01, 2025. [Online]. Available: https://ourworldindata.org/plastic-pollution
[229] A. Jenish, 'Key Players and Positions in the Global Plastic Treaty Negotiations', Earth.Org. Accessed: Jul. 01, 2025. [Online]. Available: https://earth.org/key-players-and-positions-in-the-global-plastic-treaty-negotiations/
[230] D. Main, 'Microplastics are infiltrating brain tissue, studies show: "There's nowhere left untouched"', *The Guardian*, Aug. 21, 2024. Accessed: Aug. 24, 2024. [Online]. Available: https://www.theguardian.com/environment/article/2024/aug/21/microplastics-brain-pollution-health
[231] M. Kozlov, 'Your brain is full of microplastics: are they harming you?', *Nature*, vol. 638, no. 8050, pp. 311–313, Feb. 2025, doi: 10.1038/d41586-025-00405-8.

[232] 'Microplastics in the bloodstream can induce cerebral thrombosis by causing cell obstruction and lead to neurobehavioral abnormalities | Science Advances'. Accessed: Jun. 02, 2025. [Online]. Available: https://www.science.org/doi/10.1126/sciadv.adr8243
[233] B. J. Seewoo *et al.*, 'How do plastics, including microplastics and plastic-associated chemicals, affect human health?', *Nat. Med.*, vol. 30, no. 11, pp. 3036–3037, Nov. 2024, doi: 10.1038/s41591-024-03287-x.
[234] J. Wunsh and Charron, Adian, 'Babies vs Plastics.'.
[235] S. M. Schüssler-Fiorenza Rose *et al.*, 'A longitudinal big data approach for precision health', *Nat. Med.*, vol. 25, no. 5, pp. 792–804, May 2019, doi: 10.1038/s41591-019-0414-6.
[236] B. G. Nordestgaard and A. Langsted, 'Lipoprotein(a) and cardiovascular disease', *The Lancet*, vol. 0, no. 0, Sep. 2024, doi: 10.1016/S0140-6736(24)01308-4.
[237] '2022 Our Wearable Future, Part 1: What Will New Tech Look Like?', Snyder Lab. Accessed: Jun. 16, 2024. [Online]. Available: https://med.stanford.edu/snyderlab/news/2022-our-wearable-future--part-1--what-will-new-tech-look-like--.html
[238] A. D. Zahedani *et al.*, 'Digital health application integrating wearable data and behavioral patterns improves metabolic health', *Npj Digit. Med.*, vol. 6, no. 1, pp. 1–15, Nov. 2023, doi: 10.1038/s41746-023-00956-y.
[239] 'Continuous glucose monitors on the rise after FDA approval'. Accessed: Jun. 16, 2024. [Online]. Available: https://www.bbc.com/news/articles/c1ddk1ddme5o
[240] 'Pipeline of Stapgen products for Type 2 diabetes', Stapgen. Accessed: Jan. 28, 2025. [Online]. Available: https://stapgen.com/pipeline-3/
[241] L. F. Yang *et al.*, 'Aptamer sandwich lateral flow assay (AptaFlow) for antibody-free SARS-CoV-2 detection', *Anal. Chem.*, vol. 94, no. 20, pp. 7278–7285, May 2022, doi: 10.1021/acs.analchem.2c00554.
[242] 'Bad skin influences age, not just looks'. Accessed: Jan. 31, 2025. [Online]. Available: https://www.bbc.com/future/article/20230823-the-curious-ways-your-skin-shapes-your-health
[243] S. W. Hwang, M. Atia, R. Nisenbaum, D. E. Pare, and S. Joordens, 'Is Looking Older than One's Actual Age a Sign of Poor Health?', *J. Gen. Intern. Med.*, vol. 26, no. 2, pp. 136–141, Feb. 2011, doi: 10.1007/s11606-010-1537-0.
[244] L. Ye *et al.*, 'A topical emollient mitigates the progression of cognitive impairment in the elderly: a randomized, open-label pilot trial', *J. Eur. Acad. Dermatol. Venereol. JEADV*, vol. 36, no. 8, pp. 1382–1388, Aug. 2022, doi: 10.1111/jdv.18162.

[245] H. Australia, 'About my health record – purpose, benefits and security'. Accessed: Jul. 10, 2024. [Online]. Available: https://www.healthdirect.gov.au/my-health-record
[246] E. Galkina Cleary, M. J. Jackson, E. W. Zhou, and F. D. Ledley, 'Comparison of Research Spending on New Drug Approvals by the National Institutes of Health vs the Pharmaceutical Industry, 2010-2019', *JAMA Health Forum*, vol. 4, no. 4, p. e230511, Apr. 2023, doi: 10.1001/jamahealthforum.2023.0511.
[247] 'Global Antimicrobial Resistance and Use Surveillance System (GLASS)'. Accessed: May 17, 2025. [Online]. Available: https://www.who.int/initiatives/glass
[248] 'Comprehensive assessment of the global burden of antimicrobial resistance: Trends and insights from 2000 to 2023 – American Journal of Biomedicine'. Accessed: May 17, 2025. [Online]. Available: https://ajbm.net/research-article_122632html/
[249] T. yliopisto, 'Global study finds differences in antibiotic resistance between genders', News-Medical. Accessed: Jun. 11, 2026. [Online]. Available: https://www.news-medical.net/news/20250616/Global-study-finds-differences-in-antibiotic-resistance-between-genders.aspx
[250] S. Reardon, 'Antibiotic use in farming set to soar despite drug-resistance fears', *Nature*, vol. 614, no. 7948, pp. 397–397, Feb. 2023, doi: 10.1038/d41586-023-00284-x.
[251] R. Mulchandani, Y. Wang, M. Gilbert, and T. P. V. Boeckel, 'Global trends in antimicrobial use in food-producing animals: 2020 to 2030', *PLOS Glob. Public Health*, vol. 3, no. 2, p. e0001305, Feb. 2023, doi: 10.1371/journal.pgph.0001305.
[252] Z. Zhang, S. L. Jackson, C. Gillespie, R. Merritt, and Q. Yang, 'Depressive Symptoms and Mortality Among US Adults', *JAMA Netw. Open*, vol. 6, no. 10, p. e2337011, Oct. 2023, doi: 10.1001/jamanetworkopen.2023.37011.
[253] J. Jureidini, 'Why are so many Australians taking antidepressants?', The Conversation. Accessed: May 16, 2025. [Online]. Available: http://theconversation.com/why-are-so-many-australians-taking-antidepressants-221857
[254] 'AI can make healthcare more accurate, accessible, and sustainable', World Economic Forum. Accessed: May 07, 2024. [Online]. Available: https://www.weforum.org/agenda/2023/06/emerging-tech-like-ai-are-poised-to-make-healthcare-more-accurate-accessible-and-sustainable/
[255] P. J. Bones and A. P. H. Butler, 'Eigen analysis for classifying chest x-ray images', in *Image Reconstruction from Incomplete Data III*, SPIE, Oct. 2004, pp. 88–96. doi: 10.1117/12.565175.

[256] N. Kim, 'An open AI model could help medical experts to interpret chest X-rays', *Nature*, Jun. 2025, doi: 10.1038/d41586-025-01525-x.
[257] K. Saab *et al.*, 'Capabilities of Gemini Models in Medicine', May 01, 2024, *arXiv*: arXiv:2404.18416. doi: 10.48550/arXiv.2404.18416.
[258] '(PDF) Towards Expert-Level Medical Question Answering with Large Language Models'. Accessed: May 07, 2024. [Online]. Available: https://www.researchgate.net/publication/370814210_Towards_Expert-Level_Medical_Question_Answering_with_Large_Language_Models
[259] 'The Path to Medical Superintelligence', Microsoft AI. Accessed: Jul. 02, 2025. [Online]. Available: https://microsoft.ai/new/the-path-to-medical-superintelligence/
[260] 'Genomics and Personalized Medicine: What Everyone Needs to Know® - Snyder, Michael | 9780190234768 | Amazon.com.au | Books'. Accessed: May 08, 2024. [Online]. Available: https://www.amazon.com.au/Genomics-Personalized-Medicine-Everyone-Needs/dp/0190234768
[261] 'Opyl - Clinical Trial Recruitment and Social Media Insight', Opyl. Accessed: May 12, 2024. [Online]. Available: https://opyl.ai/
[262] 'AI-Powered Clinical Trials - TrialKey'. Accessed: May 12, 2024. [Online]. Available: https://trialkey.ai/
[263] T. Schneider, 'AI-assisted breast-cancer screening may reduce unnecessary testing', Washington University School of Medicine in St. Louis. Accessed: Jul. 11, 2024. [Online]. Available: https://medicine.wustl.edu/news/ai-assisted-breast-cancer-screening-may-reduce-unnecessary-testing/
[264] I. Sample and I. S. S. editor, 'Vaccines to treat cancer possible by 2030, say BioNTech founders', *The Guardian*, Oct. 16, 2022. Accessed: Jul. 10, 2024. [Online]. Available: https://www.theguardian.com/society/2022/oct/16/vaccines-to-treat-cancer-possible-by-2030-say-biontech-founders
[265] 'Major inflammatory bowel disease cause identified – and treated', New Atlas. Accessed: Jun. 12, 2024. [Online]. Available: https://newatlas.com/medical/inflammatory-bowel-disease-ets2/
[266] C. Stankey *et al.*, 'A disease-associated gene desert directs macrophage inflammation through ETS2', *Nature*, Jun. 2024, doi: 10.1038/s41586-024-07501-1.
[267] 'Muscular dystrophy-reversing gene therapy: Human trials in 2 years'. Accessed: Jul. 23, 2024. [Online]. Available: https://newatlas.com/medical/muscular-dystrophy-gene-therapy-humans-2-years/

[268] 'Split intein-mediated protein trans-splicing to express large dystrophins | Nature'. Accessed: Jul. 23, 2024. [Online]. Available: https://www.nature.com/articles/s41586-024-07710-8
[269] B. Tran *et al.*, 'Global Evolution of Research in Artificial Intelligence in Health and Medicine: A Bibliometric Study', *J. Clin. Med.*, vol. 8, Mar. 2019, doi: 10.3390/jcm8030360.
[270] CSIRO, 'CSIRO report highlights "extraordinary era" of AI in healthcare'. Accessed: Jul. 23, 2024. [Online]. Available: https://www.csiro.au/en/news/All/News/2024/March/AI-Trends-for-Healthcare-CSIRO-report-highlights-extraordinary-era-of-AI-in-healthcare
[271] 'Vaccine-preventable Diseases'. Accessed: May 16, 2025. [Online]. Available: https://www.who.int/teams/immunization-vaccines-and-biologicals/diseases
[272] 'Child vaccination in Australia is falling', Grattan Institute. Accessed: May 16, 2025. [Online]. Available: https://grattan.edu.au/news/child-vaccination-in-australia-is-falling/
[273] R. F. Sugiura Lauren J. Young, Fonda Mwangi, Alex, 'How a West Texas Outbreak Threatens Measles Elimination Status', Scientific American. Accessed: May 16, 2025. [Online]. Available: https://www.scientificamerican.com/podcast/episode/west-texas-measles-cases-threaten-elimination-status-in-the-u-s-heres-why/
[274] 'European Region reports highest number of measles cases in more than 25 years – UNICEF, WHO/Europe'. Accessed: May 16, 2025. [Online]. Available: https://www.unicef.org/press-releases/european-region-reports-highest-number-measles-cases-more-25-years-unicef-whoeurope
[275] 'Spinal cord injury'. Accessed: May 22, 2025. [Online]. Available: https://www.who.int/news-room/fact-sheets/detail/spinal-cord-injury
[276] E. Musk, 'An Integrated Brain-Machine Interface Platform With Thousands of Channels', *J. Med. Internet Res.*, vol. 21, no. 10, p. e16194, Oct. 2019, doi: 10.2196/16194.
[277] L. Leffer, 'What It's like to Live with Elon Musk's Neuralink Brain Chip', Scientific American. Accessed: May 22, 2025. [Online]. Available: https://www.scientificamerican.com/article/neuralinks-first-user-describes-life-with-elon-musks-brain-chip/
[278] H. Magazine, 'Wounded Wilderness: The Exxon Valdez Oil Spill 30 Years Later', Hakai Magazine. Accessed: Jul. 20, 2024. [Online]. Available: https://hakaimagazine.com/news/wounded-wilderness-the-exxon-valdez-oil-spill-30-years-later/
[279] 'R&D as a percentage of GDP by country 2022', Statista. Accessed: May 05, 2024. [Online]. Available:

https://www.statista.com/statistics/732269/worldwide-research-and-development-share-of-gdp-top-countries/

[280] R. Merritt, 'What Is a Transformer Model?', NVIDIA Blog. Accessed: May 07, 2024. [Online]. Available: https://blogs.nvidia.com/blog/what-is-a-transformer-model/

[281] P. Dar, 'A Technical Overview of AI & ML (NLP, Computer Vision, Reinforcement Learning) in 2018 & Trends for 2019', Analytics Vidhya. Accessed: Jan. 23, 2025. [Online]. Available: https://www.analyticsvidhya.com/blog/2018/12/key-breakthroughs-ai-ml-2018-trends-2019/